Apologia

During some twenty years in industry, I have had to read a large number of management books of one sort and another—many connected with my work, but many outside it.

Among all of these there have been only a handful which have kept my interest and made a lasting impression; other people's experience must be much the same. So, when asked to take over the editorship of this series, I was glad of the opportunity to try to do what I could to add to the number and to introduce some of my favourites to a wider audience.

I have often felt that the man who most needs to learn is exactly the man for whom very little is written. The new manager is usually among the keenest to know and understand what goes on—but he is very poorly treated by the usual run of academic dispute, of advice from adepts to experienced practitioners, and of superficial introductions to various subjects. The new manager wants very little of these; if they are all that is offered, he is likely to cry 'to the devil with both your houses', and to carry on his work by whatever light his nature casts. A loss, not only to him, but to business as a whole.

The books in this series are all specifically addressed to the new manager and his problems. The most immediate of these problems are those of getting on with colleagues in other functions—in all but the smallest of companies the invisible walls which surround functional empires tend to be tall and impenetrable; and arguments on who does what on the imperial frontiers are not of concern only to the trade unions. So, one task of the Library will be to describe other people's jobs—what goes on behind some of the closed doors along the corridor. There are problems there, too, and the apparently contentious conflicts which come through the door are usually due to the logic of the problems and not to backstairs intrigue by those on the other side, nor even to ordinary cussedness.

Quite apart from this effort to bring understanding and so promote co-operation on the business scene, there are at least two other reasons why an awareness of other people's jobs will be useful. First, a broader outlook should counter parochialism, and plant a few doubts on whether super-efficiency at one's own job regardless of anything else is ever the right thing in the general interest.

More importantly, an understanding of the problems of all the functions is a very good key to the door leading to general management—if not the only one. It is, really, only generalists like accountants who have to have this essential equipment of a general manager as part of their work; it might be a good thing for the specialists to indulge in some extra-curricular activity, to acquire the qualifications too—and it will probably be in this country's best interests also, because there is a lot of talent outside accountancy which tends to wither in a subordinate role.

The other set of problems is how best to deal with the job in hand. To the beginner, there appears to be an immense tool-bag, filled to overflowing with a vast range of techniques, each with its own vociferous crowd proclaiming its merits—and with an equally large crowd busily denigrating it. But it is surprising how far one can get with no more than a hammer, a screwdriver, a pair of pliers, a vice, a saw, some chisels, a drill, a plane, and a few assorted files.

There are only a few vital techniques which, when used with common sense, are all that is needed to solve most problems. So, the other task of the Library will be to describe these fundamental techniques, to show what they can do, what they cannot do, what dangers attend their use and what has to be done if something more complex is needed.

Finally, since all business is a story about people, the books will not allow themselves to forget them, with all their complexities—which does not mean an unpalatable dose of the social sciences. What it does mean is that any conclusion amply proved in practice will be incorporated into the Library's outlook, whether it comes from the physical or the social sciences, or from some more commonplace source. This is very far from meaning that only the established truth will be revealed; much of the Library will be concerned with fresh insights, and will let the controversy fall where it will.

No one, not even the authors of books in the Library (and certainly not its general editor), knows more than a small fraction of the truth. Anyone, particularly those at the start of their careers in management, can have good and workable new ideas, given the stimulus.

About this book

Among the techniques which operational research has fathered, network analysis with the critical path has had an immediate success.

The New Manager's Library
General Editor: D. R. C. Halford, O.B.E.

Critical Path Planning
a practical guide

The New Manager's Library

The Practice of General Management
catering applications
David A. Fearn

Profit Management and Control
Fred V. Gardner

Direct Standard Costs
for decision making and control
Wilmer Wright

Investment in Innovation
C. F. Carter and B. R. Williams

It is now firmly established as the best way yet invented for the planning and control of large projects of various sorts, where many and diverse activities have to be brought together.

One of the things which makes it so appealing is the apparent simplicity of the first task: to list all the activities and put them in their logical sequence. Several hours (days, or even weeks) later, when the job is done and loops have been abolished (*A* comes before *B* which precedes *C* which is a prerequisite for *A*), we will know very much more about the project than when we started.

Of course it gets harder as we go further into the problem—to the allocation of resources. But the appeal of a good, challenging and surely not insoluble puzzle remains; this is one technique where the standard exhortation to plan and estimate right, stands a very good chance of being followed.

D. R. C. Halford
General Editor
January 1971

Critical Path Planning
a practical guide

K. M. SMITH
Planning and Development Manager
British Transport Hotels Ltd

Macdonald . London

First published in 1965 by the British Institute of Management under the title
A Practical Guide to Network Planning

Published 1969 by Management Publications Ltd
for the British Institute of Management
This revised edition first published in 1971 under the title
Critical Path Planning: a practical guide in 1971
in hardback by
Management Publications Ltd
and in paperback by
Macdonald & Co. (Publishers) Ltd
49–50 Poland Street
London, W.1

SBN 356 03488 7

PRINTED AND BOUND IN ENGLAND BY
HAZELL WATSON AND VINEY LTD
AYLESBURY, BUCKS

Contents

Preface to the first edition

In November 1963 I was asked to prepare and present, jointly with Mr. D. Williams of Operational Research Department, a series of three-day courses of instruction in Network Planning Techniques* for Senior Officers at the British Railways Board Headquarters and in the Regions of British Railways. This book, written at the request of the British Institute of Management, is based largely on the material presented on these courses. I am indebted to the British Railways Board for permission to publish it, particularly the example and diagrams in Chapter 8.

This is a book of instruction. It outlines the simpler aspects of Network Planning Techniques and is intended for newcomers who wish to learn to use these techniques in planning projects of various kinds. Theory has been cut to a minimum and the mechanics of the techniques are presented in Chapters 2–6 as a simple drill which can be readily mastered. In Chapter 8 an application to a modest civil engineering project, without restort to the use of a computer, is explained.

The reader need have no knowledge of mathematics beyond the ability to handle simple arithmetical calculations. A knowledge of some simple statistical principles is, however, required in order to understand the explanation in Chapter 7† of the principles of the special Network Planning system known as PERT (Programme Evaluation and Review Technique). Nor need the reader have any previous knowledge of computers. A simple explanation of their use in Network Planning is given in Chapter 5.

Although the illustrations and examples are taken mainly from the civil engineering field, Network Planning Techniques have already proved their worth in planning projects in many other areas of engineering, and indeed elsewhere. Extensions of the technique for the more efficient allocation of resources are still being developed and the solution of these problems will be of tremendous value in future

* See Author's note—terminology, page xiv.

† Placed in an appendix in the second edition.

project planning, progressing, and control as project size and complexity continue to increase.

The techniques themselves will not, however, solve all planning problems. No matter how good the techniques are much has still to be done in building up experience in their use in different circumstances and in different types of work.

For many helpful suggestions on the content and presentation of this book, I am indebted to Mr. David Williams* and Miss Gail Boden† of the Operational Research Department, British Railways Board, and to Mr. A. G. Kentridge‡ of British Railways, Eastern Region.

K. M. Smith
Marylebone, London
June 1965

* Now Management Consultant, Planned Warehousing Ltd.

† Now Mrs Gail Thornley, Department of Operational Research, University of Lancaster.

‡ Now Executive Director, Planning, British Rail Headquarters.

Preface to the second edition

The second edition remains essentially a book of instruction. The section on the matrix method of analysis of the network has been removed from Chapter 5 as being of little practical value. The chapter on the principles of PERT and CPM (Chapter 7 in the original edition) has no real relevance to the subject matter or purpose of this book. It has therefore been relegated to an appendix where those of an academic turn of mind can acquire, in a very simplified form, an understanding of the basis of these early systems.

A new Chapter 7 under the title of 'Monitoring and Control' gives what is of much more relevance and practical value to the practising manager, namely an insight into some of the practical problems which have to be faced in using these techniques. Also included is a brief outline of the computer aids available for cost analysis and resource planning which have been more fully developed since the first edition.

I am particularly indebted to Mr. D. R. C. Halford, O.B.E., General Editor, New Manager's Library, and to Mr. Michael Justice, Editor, Books Division, Management Publications Ltd., for their many helpful suggestions in the preparation of this second edition.

K. M. Smith
St. Pancras, London
October 1970

Author's note—terminology

There has never been universal agreement on a generic term for the subject of this book. Critical Path Analysis (CPA) and Programme Evaluation and Review Technique (PERT) are popular terms but both are also names given to particular variations of the basic network technique. The name 'Network Planning' was used in the title of the original edition of this book because it signified the use of *networks* for *planning*. Because it is more widely known, the name 'Critical Path Planning' is now used in the title of this edition although in the author's view is gives an undesirable emphasis to the critical path which is but one aspect of the whole technique. The term 'Network Planning' is still used throughout this text to refer to the basic technique.

K. M. Smith
St. Pancras, London
October 1970

One Historical background

Having recognized the limitations of traditional methods of planning industrial projects, two teams of consultants in America applied themselves independently, in 1957, to the task of devising a better system of project planning, using the basic principle known to mathematicians as the theory of networks.

One of these projects was the design and development of the Polaris missile. With the objective of meeting a specified end date for the project two years in advance of the earliest possible date predicted by traditional planning methods, Booz, Allen and Hamilton, consultants, together with the United States Naval Department, devised the now well-known system of Network Planning,* which was given the name PERT (Programme Evaluation and Review Technique).

The second of these projects embraced the work involved in a chemical plant overhaul. It was initiated by the Du Pont chemical company in America with the object of reducing the length of time a chemical plant would be out of commission during a major overhaul. The Remington Rand Corporation provided the computer know-how and together with Du Pont they devised the system of Network Planning known as CPM (Critical Path Method).

Both these original systems of Network Planning are briefly described in an appendix at the end of this book. They were evolved by mathematicians and computer experts, and both proved to be highly successful in practice for the purpose for which they were designed. Their value in planning projects of practically any kind and with any objectives in view was quickly recognized, and gave rise to a number of further variants of the same basic technique, for example PEP (Programme Evaluation Procedure), COPAC (Critical Operating Production Allocation Control) and MAP (Manpower Allocation Procedure).

The predominant issue in using these early Network Planning Techniques was minimization of time. Later developments have in-

* See Author's note—terminology, page xiv.

troduced as objectives the determination and minimization of project cost and the allocation and levelling of resources (manpower, money, machinery), and so on, but the achievement of these objectives is much more difficult. A means of automatically optimizing all the conflicting objectives in a project has yet to be devised.

Network Planning Techniques are now extensively employed in Great Britain, and the simpler forms of these techniques are used for planning a diversity of projects in many major building and construction companies, oil companies, in British Railways and elsewhere. Many computer firms and management consultants offer assistance and facilities for this type of work. Some Civil Service departments in this country are already following the American practice of making it obligatory to submit a network analysis in support of a quotation for a government contract, and this practice is likely to spread in the future when the power of these techniques is fully recognized.

Two **Basic concepts**

The need for planning

There has been a growing awareness in industry in recent years of the need for better planning, of the need to look ahead and anticipate snags before they arise. Not only has this fact been recognized at the level of the national economy, but senior, middle and junior managers in industry, faced with the ever increasing complexities of their work, have recognized the need for some aid to controlling and co-ordinating the many diverse operations involved in any modern industrial project. Network Planning Techniques fulfil at least a part of this need.

Scope for network planning techniques

Network Planning Techniques have their application in project-type work only, that is to say, in planning projects which have precisely defined start and end points. This does not however mean that the use of these techniques is confined to civil engineering and construction projects. Indeed they have many applications in engineering design, maintenance and manufacture, in major plant overhauls, in launching new products on to the market, and even in the production of theatre shows. It is necessary at the outset only to define the project by specifying precisely its start and end points.

Network Planning Techniques are most successful where the problem is one of co-ordinating and controlling a large number of concurrent activities directed towards the same goal. It is possible to recognize many situations in which co-ordination of contractors, departments or tradesmen and machines is required. For instance, in the design and development of a prototype aircraft, the design, construction and commissioning of a new chemical plant, or in any sizeable building or civil engineering project, Network Planning provides an excellent vehicle for co-ordination and control at all levels in the organization.

In projects where the constituent activities must of necessity take place consecutively however, Network Planning Techniques will have little to offer.

It should also be emphasized that these techniques cannot in any way help in deciding what the project should be, whether for instance a pipeline or a bucket elevator should be installed to transport intermediate materials from one part of a factory to another or whether a new hotel should have 200 or 500 bedrooms. Network Planning Techniques can only help in planning *how* to carry out *what* it has already been decided to do.

Limitations of bar chart planning

The traditional method of project planning normally involves preparing a bar chart schedule or programme. This is done by placing bars representing the major blocks of work in the project against a time scale in a way which previous experience of similar projects has shown to be feasible. For example the installation of a large shed in the garden might be scheduled on the bar chart in Figure 2.1.

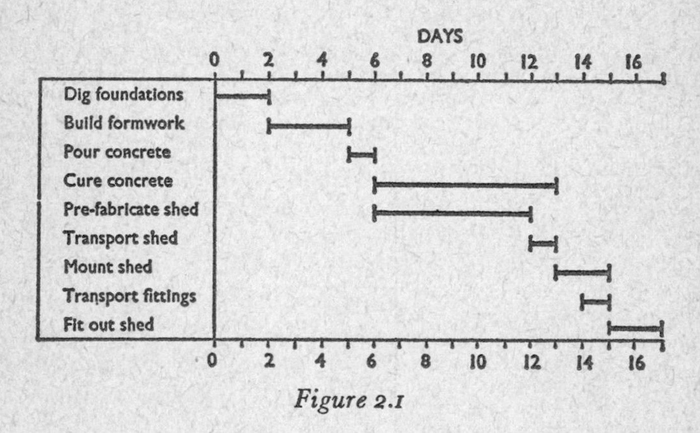

Figure 2.1

This method of programming has limitations in that, in a large and complex project, it is too cumbersome to enable the activities in the project to be considered in sufficient detail, and to show the vital interactions between these activities. It does not therefore provide a satisfactory means of predicting far enough in advance the effects of inevitable snags and delays, to enable corrective action to be taken in time. And it gives no indication as to which activities should have priority for resources. The result is that the project tends to progress from crisis to crisis, with the attendant cost of inefficient utilization of resources, and failure to meet promised completion dates.

The advantages of network planning

The preparation of bar charts is merely a scheduling or programming operation. Network Planning takes into account the need for planning in its true sense of first placing in their logical order the activities in the project. The network showing the logical order for the activities in the installation of a large shed in the garden might be as in the network diagram in Figure 2.2.

Whether planning the activities of contractors, departments or tradesmen, the technique enforces a discipline which automatically

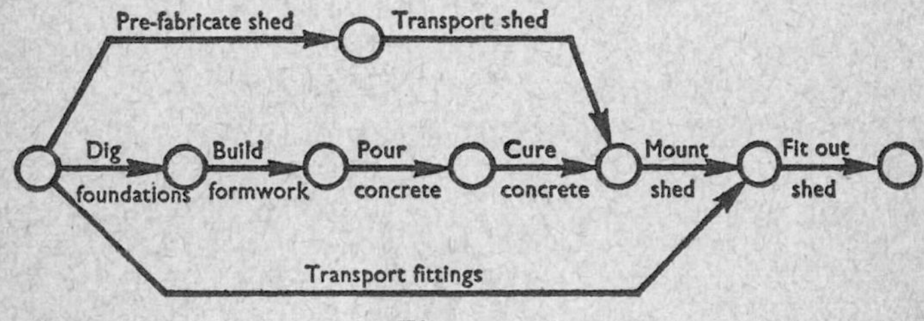

Figure 2.2

shows how each activitity depends on the others. A system is thereby provided for monitoring the progress of the project, for forecasting the effects of snags on the project as a whole, and for deciding which activities should have priority for resources. The action necessary to avoid a crisis can be taken in plenty of time, resources are better utilized, completion dates can be predicted with confidence and, what is more important, achieved.

The mechanics of how these planning techniques operate is the subject of the succeeding chapters.

The basic steps of the technique

There are four basic steps in network planning :

1 The preparation of a network of arrows in which each arrow represents one activity, job or task in the project, and the positions of these arrows in the network represent the logical order in which the activities must take place (see Figure 2.2).

2 The estimation for each activity of the duration and if necessary the direct costs and resources required.

3 The analysis of the network: to identify the activities which are critical in the sense that they govern the over-all length of the project; and to provide data on the range of possible starting times for each activity and perhaps on the level of demand for resources throughout the project.
4 The preparation of a schedule or programme possible in bar chart form, using the information provided by the analysis of the network to help achieve the best programme possible in the circumstances (see Figure 2.1).

These four steps emphasize the distinction between planning and scheduling. The preparation and analysis of the network represents the planning function, providing information for the separate and distinct process of scheduling.

Updating the schedule may involve a repeat of all four steps. A change in the method or logic of the project, or a change in the estimate of the duration of an activity, can necessitate a revision and re-analysis of the network and the preparation of a new schedule.

The next four chapters are devoted to considering, in detail, each of these four steps in the procedure.

Three **Network diagrams**

Introduction

The fundamental basis of Network Planning Techniques is the preparation of a network diagram. In it each activity or task in the project is represented by an arrow. The positions of these arrows in the network represent the logical order in which the activities must be performed.

Having prepared the network diagram it is necessary to mark and number the points at which arrows meet. These points are called nodes or events and the numbers they are given act as a basis for describing and analysing the network.

The first two sections of this chapter deal with the preparation of the network diagram and the addition to it of numbered events. A third section deals with the managerial problems of setting up an organization to plan a project using Network Planning Techniques.

Section 1 **Activities and logic**

Activities represented by arrows

The network diagram is the foundation stone of all techniques of network planning. It represents in schematic form the over-all plan for the project. In it each of the constituent activities or tasks is represented by an arrow in which the direction of time flow is from tail to head, and the positions of these arrows in the network represent the logical order or sequence in which the activities must be performed. Activities normally require time, money, manpower and other resources for their completion.

Because arrows are used to represent activities, the diagram is often referred to as an arrow diagram. An illustration of a simple network diagram has already been given in Figure 2.2.

Placing the arrows in the network diagram

Without considering for the moment how a project is broken down into its constituent activities, let it be assumed that this can and has been done and that the start and end points of the project have been clearly defined. The network diagram must now be constructed step by step, placing each arrow representing an activity in the diagram according to its interdependency on the other activities in the project. This is done quite simply by considering each activity in turn and asking two questions:

(i) What activities must precede and be completed before this one can start?

(ii) What activities must follow this one?

In answering these questions it is important to disregard any restrictions which may ultimately be imposed on the performance of these activities in the form of artificial target dates or limitations on the availability of resources. What is required is to consider 'what activities *must* precede . . .' and 'what activities *must* follow . . .' from the purely technological point of view. For example when painting a small wood surface the application of the priming coat must precede and be completed before the application of the undercoat, and application of the finish coat must follow the application of the undercoat. The three activities: prime, undercoat and finish coat must take place in that order and in no other order. They must therefore be represented in the network diagram thus:

prime ⟶ undercoat ⟶ finish coat ⟶

Similarly for setting a pole in the ground, dig hole, set pole and backfill are three activities the order of which is quite irreversible:

dig hole ⟶ set pole ⟶ backfill ⟶

In the slightly more complicated example of building a house it is necessary to complete the electrical carcase, the gas piping and internal plumbing before fixing plaster boards and plastering, and plastering must be followed by fixing electrical fittings, plumber's fittings, doors, and so on. This would be represented on the network diagram thus:

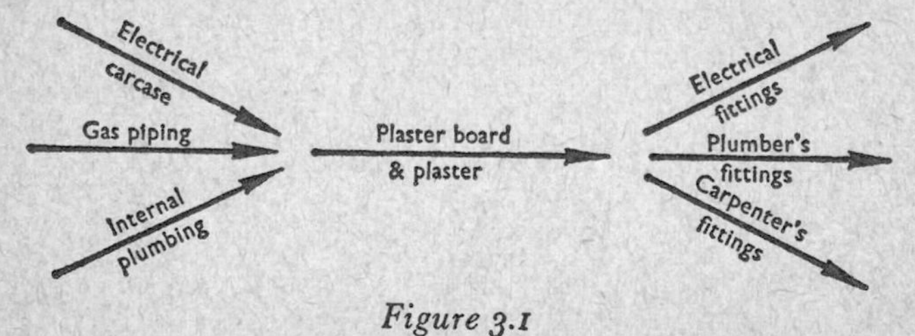

Figure 3.1

The process of drawing the network diagram is purely and simply one of deciding the technological sequence of the activities. Neither the individual arrows nor the general configuration of the network need follow any scale, time or otherwise. The only object at this stage is to represent in schematic form the order in which the activities *must* take place by virtue of their technological interdependencies.

As a final confirmation that an arrow has been placed in the correct position in the network a third question should be asked:

(iii) Which activities can be performed concurrently with this one?

This ensures that no links or interdependencies with other activities have been overlooked.

Exercises

1 Draw an arrow diagram to represent the following interdependencies only:

Job A immediately follows jobs B and C; jobs B and E immediately follow job D.

2 Draw an arrow diagram to represent the following interdependencies only:

Jobs L and M cannot start until K is complete; job L must be completed before job N can start; job M must be followed by job Q; both N and Q must be completed before P can start.

(Answers to all the exercises are given at the end of the book.)

Dummy activities—preservation of logic

It is often necessary to introduce into the network diagram dummy activities in the form of dotted arrows in order to preserve the logical sequences of activities. For example, if a car is to have a replacement engine fitted the replacement engine must be assembled and the old engine removed from the car before the replacement can be fitted; the old engine can, however, be stripped as soon as it is out of the car without necessarily waiting for the replacement engine to be assembled. Of the following two diagrams therefore, (*a*) does not represent the facts as stated but (*b*) does:

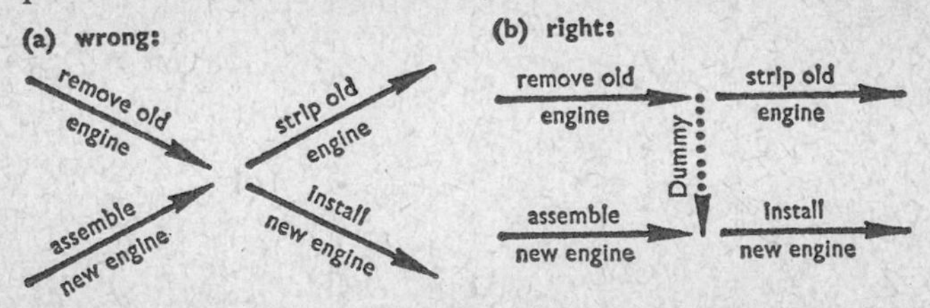

Figure 3.2

Whereas real activities consume both time and resources for their accomplishment dummy activities consume neither time nor resources. Dummy activities are therefore treated in all respects as activities of zero duration.

Overlapping activities

The situation often arises where activities can partially overlap. They can neither start at the same time nor does one have to await the completion of the other. For example where a long length of pipe is to be laid in the ground the following straightforward arrow diagram will not be a true representation of the situation :

trench	lay pipe	weld joints	backfill
⟶	⟶	⟶	⟶
200 yards	200 yards	200 yards	200 yards

It will be possible in fact to start laying pipe before trenching is completed, the start of welding need not await the completion of pipe laying, and so on.

What is required to make the arrow diagram reasonably representative of the true situation is to show the bulk of the overlapping activities proceeding simultaneously while showing the minimum stagger at the beginning and end.

For example trenching and laying pipe can be shown overlapping thus :

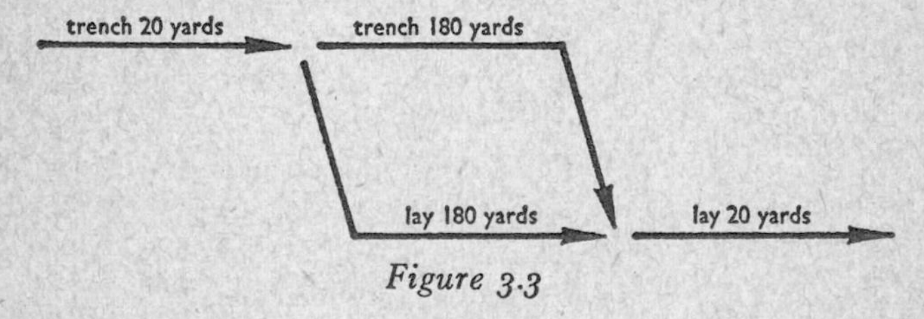

Figure 3.3

This shows that all trenching must be completed before the last twenty yards of pipe is laid, an indication that the laying operation cannot overtake the trenching operation. Strictly speaking of course, to make sure that laying cannot overtake trenching in the arrow diagram, both operations could be broken down into a larger number of smaller parts instead of just two but this would produce a degree of detail which is hardly likely to be of practical value to the user of the network (see page 15 : Scale of detail—size and choice of activities).

What is required is a reasonable representation of how the operations will be carried out in practice, showing the 'lead' stagger for the first operation and the 'lag' for the last operation. If the time per unit length differs from one operation to another this will be represented in the way the operations are broken into parts, and the 'lead' stagger will of course differ from the 'lag'.

The diagram of overlapping arrows might take the form shown in Figure 3.4.

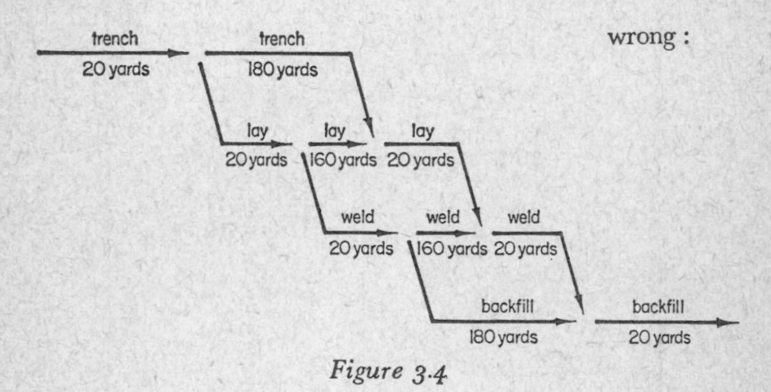

Figure 3.4

This diagram is not yet quite logical however because it suggests for example that it would be possible to backfill 180 yards having trenched, laid and welded only the first twenty yards.

It is necessary also to avoid building into the arrow diagram irrelevant activity dependencies as for example :

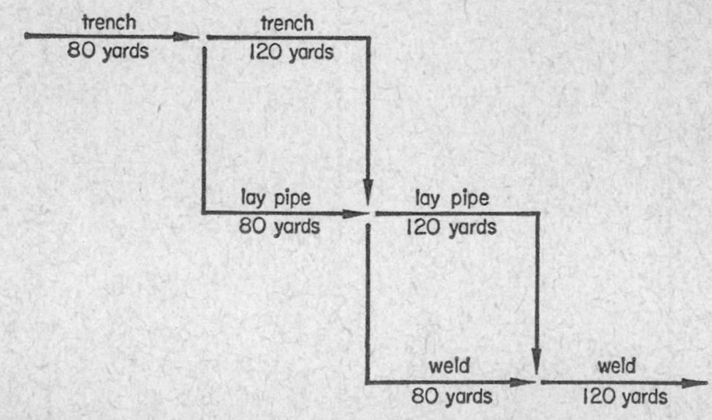

Figure 3.5

This suggests that 'weld eighty yards' depends on 'trench 120 yards' which is clearly not the case. The insertion of a dummy arrow will rectify this:

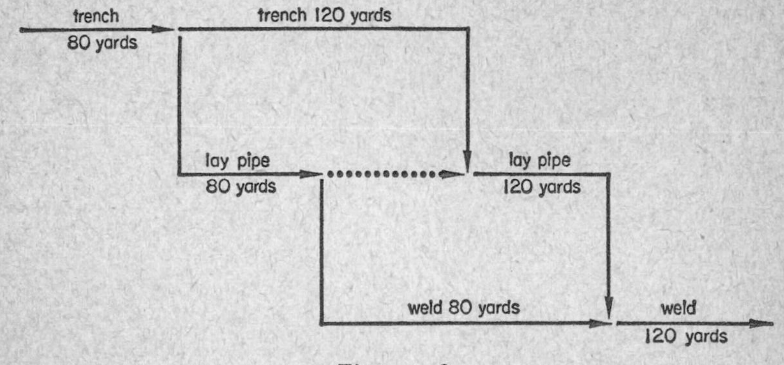

Figure 3.6

Depending on the number of operations which overlap and the number of parts into which each is divided, and employing dummies to preserve the logic, the situation might be represented by a diagram similar to that in Figure 3.7.

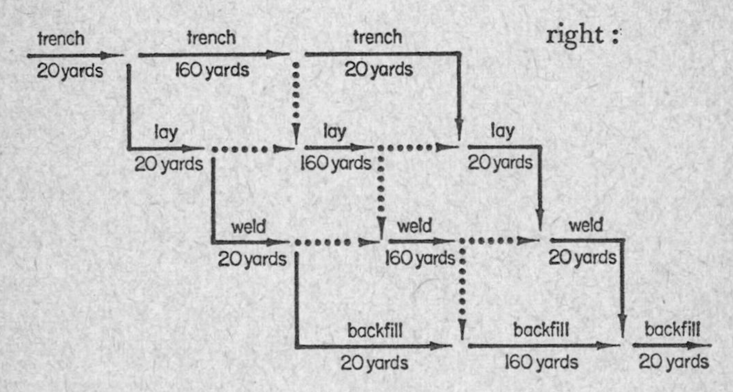

Figure 3.7

This diagram correctly indicates that the four operations could be carried out as a complete job independently in each of the sections of trench or alternatively that the sections could be linked to provide continuous work for each gang. Maximum flexibility is thus retained for scheduling these operations.

In practice this type of representation would most often be employed in refining a network to make it more closely represent the practical outcome (see page 44 : Refining the network to meet management objectives). The activities concerned may not have been overlapped in the first drawn network but when they are found to be on the critical path there will be a need to refine the method of representing these activities. This is done by overlapping them, thus reducing the time given to them in the network. When the 'lead' stagger (twenty yards in the illustration) is the minimum practical distance between two consecutive gangs then the diagram represents the maximum overlap and the minimum time for the whole job. It is therefore the closest representation of the actual outcome, which can be achieved in practice although it may not be necessary in the event to take advantage of the maximum overlap.

It may appear that this arrow diagram represents a limitation in resources in that, for example, trenching in the three sectional lengths must be consecutive. Why, it may be asked, can three separate gangs not start in the three sections simultaneously? This indeed could and would be done if the over-all time for the four operations turned out to be too long if carried out according to the arrow diagram in Figure 3.7. Indeed, if the lead and lag times are so small as to be insignificant then the four activities could be shown as parallel activities in the network without undue loss of accuracy. This is however to look at this particular point in a slightly different context. The arrow diagram in Figure 3.7 is an attempt to represent the way these activities could overlap in practice under 'normal' conditions. Although each operation has been split into three arrows this has been done only for convenience in showing the overlap; the three arrows still really represent one activity.

Activities which consume only time

In certain circumstances it is necessary to have in the network diagram arrows which represent activities which consume only time and no other resources. 'Allow concrete to harden' or 'allow paint to dry' are examples. Such 'activities' must be put in the network to show that they must follow the 'pour concrete' and 'paint' activities respectively, and must be completed before, for example, building on top of the concrete or applying a second coat of paint. These 'time-only' activities can influence the possible starting and finishing times of other activities in the network and must therefore take their logical place in the network. They are treated as activities of the specified duration, but requiring no resources.

Logical loops

It sometimes happens, through some fallacy in logic, that a chain of arrows in the network forms a closed loop implying that, after carrying out a number of activities, it is possible to return to the point in time from which a start was made. Clearly this is impossible. This can arise where any kind of recycling or iterative process towards a choice or decision is involved as for example in the development of a production process :

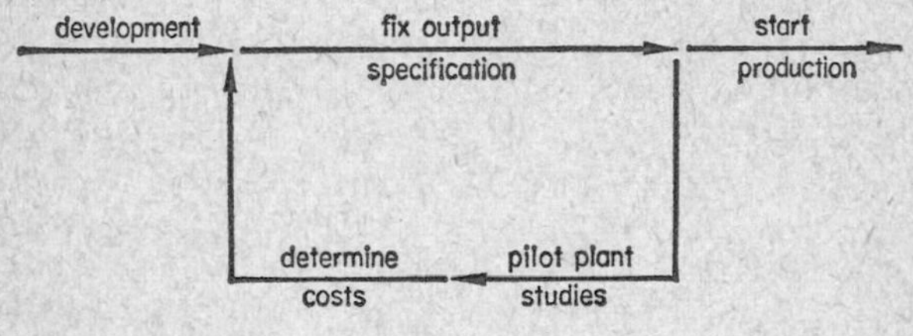

Figure 3.8

The problem can be resolved by making the choice or decision depend on all the work which may influence it :

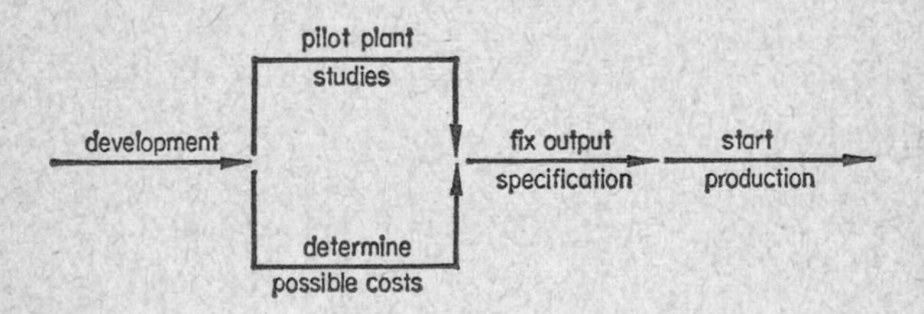

Figure 3.9

An actual project in which a 'loop' of several activities emerged is described in an article entitled 'Planning the easy office move' in the February 1968 issue of *Business Management*.

Scale of detail—size and choice of activities

The construction of the network diagram is not quite as simple as it may appear. Although in every project the activities naturally fall into logical order it is not always easy, particularly in long and complex projects, to identify the logic and describe it on a network diagram.

One difficulty is in deciding how to break the project down into its constituent activities and deciding how large or small the activities ought to be; in other words what degree of detail should be considered. There is unfortunately no general answer to this question as it depends so much on the size and complexity of the project and what objectives are envisaged. The only general guide is to consider the needs of the individuals who are going to use the network and interpret the results of its analysis. The manager of a large project will be interested in activities of not less than about a week's duration while the foreman will be interested in operations of not less than about a day's duration. The choice of activities in the network should anticipate these needs.

It is not however serious if the first choice is not always the correct one. The practical approach to drawing networks is with a pencil and rubber, and very large sheets of paper. Long arrows should be drawn in the first instance and subsequently broken down into several smaller arrows if this is found to be desirable. It is sometimes advocated that all the activities should be listed before starting to draw the arrow diagram but in practice it is generally found that only by drawing the arrow diagram is it possible to decide on how to break the project down into its constituent activities. In other words an empirical approach with the eventual needs of the user in mind, is required.

No two people will ever draw exactly the same network for the same project. The scope for differing interpretations of the size and choice of activities, and for differing logic representing different methods and procedures for doing the same job, make for an infinite variety of possible networks for the same project. What is important is that the logical disciplines involved in using network planning techniques offer an opportunity of examining traditional methods and procedures and evolving new ones (this alone can be a salutary exercise), and that the network prepared should represent a practical, if not necessarily the only, method of doing the job.

Presentation on paper

It is customary, though not necessary, to draw straight or dog-leg arrows rather than curved arrows, and it is convenient in practice to draw at least part of the arrow as a horizontal line against which a brief description of the activity can be written. This makes it easier to locate arrows more quickly than it would be referring to a coded list of activities. There is no reason why arrows should not cross.

It is also usual to work from left to right of the paper but the layout of the diagram is unimportant provided the network remains an intelligible and realistic representation of the logical order of the activities in the project.

The network diagram can be drawn in sections on several sheets of paper, confining the activities of one department to one sheet of paper if desired and linking by arrows in the normal way. The diagram as a whole will however have a single start and a single end point defined by the start and end points of the project.

Exercise

3 Draw an arrow diagram to represent the following interdependencies only:

> Jobs *A*, *B* and *C* can be carried out simultaneously after *N* has finished. *P* cannot be started until *D*, *E* and *L* have been completed. *D*, *E* and *L* can proceed at the same time but *A* must be finished before *D* can start. *L* cannot start until *A*, *B* and *C* are finished. *E* depends on *G*, *H* and *K* being completed; these can proceed concurrently but *G* cannot start until *F* is completed and *J* must be finished before *K* can start. *J*, *F* and *H* can be carried out together but cannot commence until *B* is completed. *N* is the first job, *P* the last.

Section 2 **Events and numbering**

Events in the network diagram

Events or 'nodes' must now be marked on the network diagram. Having prepared the diagram to represent the interrelationships between the activities a small circle, later to contain a number, is drawn at each point (i.e. node) in the network at which tails and heads of arrows meet. For example, the network in Figure 3.1 would have circles added as follows:

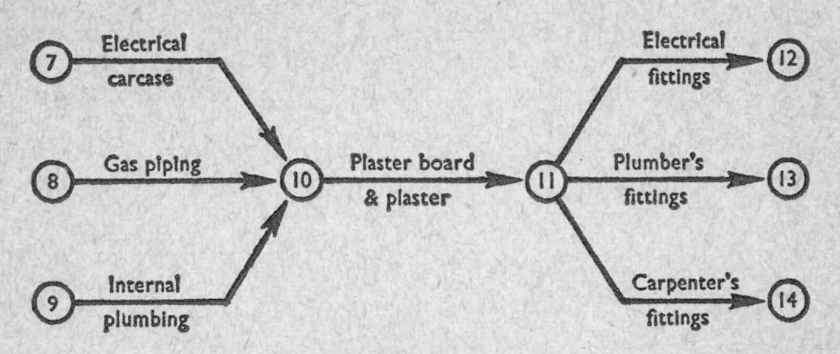

Figure 3.10

These circles mark the 'nodes' or events in the network and act as milestones through it. In other words an event signifies a point in time or stage reached in the progress of the project. Event 10 for instance is the stage reached in the project when the electric carcase, gas piping and internal plumbing are complete and plastering can be started. This is not to say that plastering must necessarily start without any time lag—no time scale has yet been introduced. It merely means that at event 10 the stage has been reached at which plastering *may* be started.

The events are numbered in such a way that, are far as possible, the number of the event at the head of each arrow is greater than the number of the event at the tail. This is done for convenience in locating particular activities or events, given their numbers, but if some of the events have acquired numbers out of sequence in the course of alterations to the network, this does not in any way affect the validity or analysis of the network.

The event-numbering system provides a means of referring to each of the activities in the network by a pair of numbers. For instance, electric carcase in the previous example can be referred to as activity 7–10, plaster board and plaster as activity 10–11, and so on. It is thus possible to describe the activities in the network and their interrelationships in purely numerical terms by assigning to each activity the appropriate pair of event numbers. This is the essential purpose of numbering events and a prerequisite to analysing the network.

Dummy activities—unique numbering system

It is possible to find in preparing a network that two or more activities can start at the same point and must finish by the same point. This means that all these activities would run between the same pair of events. For instance in the previous example, the plumber and carpenter can both start to instal fittings as soon as plastering is complete, and both must be finished before interior decoration commences. If this situation were represented by the following diagram:

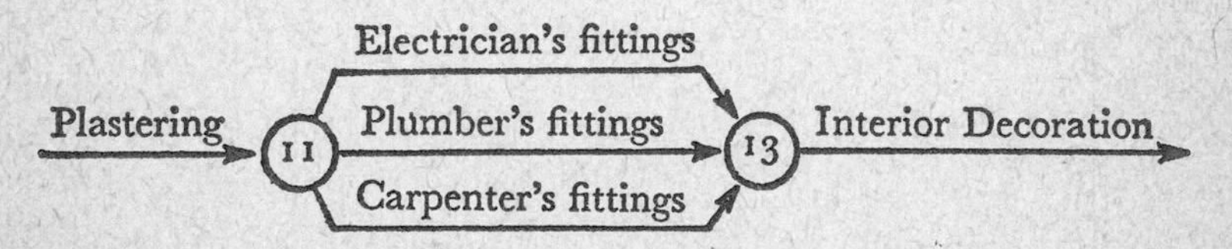

reference to activity 11–13 would be ambiguous; it would be impossible to know whether reference was being made to the electrician's, plumber's or carpenter's activity.

In order to preserve a unique numbering system in which each activity is described by a unique pair of event numbers it is necessary in this situation to introduce a dummy activity (shown as a dotted arrow) and a further event thus:

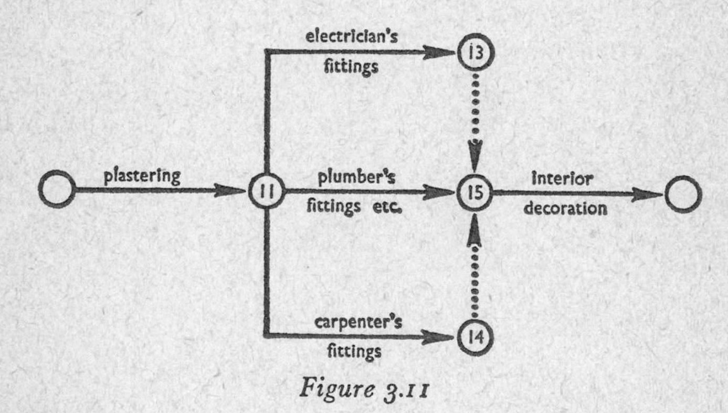

Figure 3.11

The electrician's activity is now uniquely described as activity 11–13, the carpenter's as activity 11–14 and the plumber's as activity 11–15.

The dummy activity can be treated in all respects as an activity of zero duration, and consuming no resources.

The two diagrams in Figure 3.12 represent the same logic as in Figure 3.11 although the activities are numbered differently.

The layout of the network is immaterial so long as the logic is preserved. It is preferable however to adopt the habit of inserting the dummy in a position preceding the activity in the sequence as in the lower diagram in Figure 3.12 because any free float associated with the sequence will automatically be attached to the activity rather than to the dummy (see also page 64, first new paragraph).

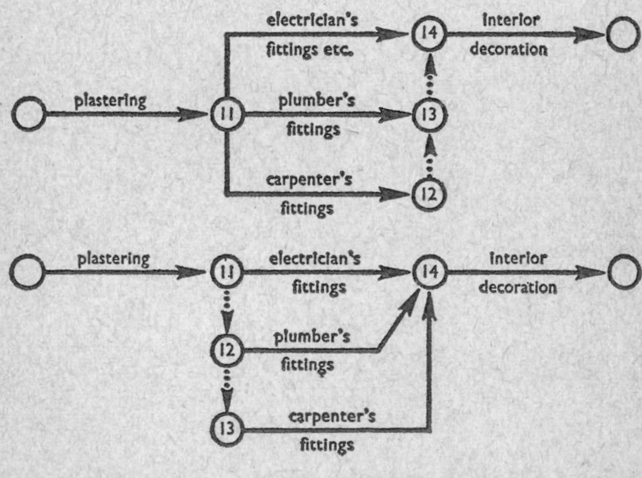

Figure 3.12

Exercises

4 Draw an arrow diagram to represent the following interdependencies only:

The starting activities *A* and *B* must both be complete before activities *C*, *D* and *E* can start. *C* and *D* must both be complete before *G* can start. *C* must also precede *F*, but not necessarily *K*, and *D* must precede *K* but not necessarily *F*. *E* is followed by *J*, *G* and *H*, and *F* by *I*, *J* and *L*. *H*, however, need precede only *I* and *J* but not necessarily *L*. *I*, *J*, *K* and *L* are all terminal activities.

5 Draw an arrow diagram to represent the following interdependencies only:

> Activities R, S and T can start immediately. R must be completed before U and W can start. X and Y must follow both S and T. V must also follow S and precede W. U, W and X must all be completed before Z can start. Y and Z are both terminal activities.

6 Link the following activities together in sequences to form an arrow diagram for the purchase, renovation and opening of a new guest-house:

(i) search for property
(ii) survey property
(iii) architect's outline plan
(iv) arrange finance
(v) solicitor's search
(vi) obtaining planning permission
(vii) architect's detailed plan
(viii) complete purchase
(ix) engage staff
(x) order furniture, etc.
(xi) seek tenders and engage contractor
(xii) renovation
(xiii) deliver furniture, etc.
(xiv) install furniture, etc.
(xv) prepare and arrange advertising
(xvi) advertise.

Summary of conventions in preparing network diagrams

It is now possible to prepare a list of the essential rules or conventions to be adhered to in preparing arrow diagrams:

1 Each arrow representing an activity starts at an event and finishes at an event.
2 The numbers of the events at the ends of an arrow must uniquely define the arrow and its position in the network.
3 The number of the event at the tail of an arrow should if possible be less than the number of the event at the head; time flow is then from the smaller to the larger number.

4 The sequences through the network must be unidirectional, i.e., there can be no closed loops or double tracking; this follows from the fact that events are points in time and once an event (point in time) has been passed it is impossible to return to it as time itself is unidirectional.
5 Dummy activities of zero duration are used in arrow diagrams to:
(*a*) preserve the logical sequence of activities
(*b*) preserve the unique activity numbering system.

The best advice on the preparation of arrow diagrams is to be equipped with a pencil, a large rubber and several large sheets of paper, and to place oneself in the shoes of the individual who will wish to use the results of the analysis of the arrow diagram—this empirical approach will help determine the size of activities to be represented in the arrow diagram.

Furthermore in the interests of speed it is bad policy to dwell excessively upon some minor point of detail in the diagram. It is in the nature of things that mistakes will be made either in the logic or in the numbering. These 'bugs' will be discovered during the analysis of the network if not before, but an analysis of the network cannot be carried out until the end of the network is reached.

In a sense the network diagram is never complete. It is a dynamic representation of the work in a project and is continually being adjusted to cater for changes in decision in regard to methods, unavoidable and unforseeable changes in circumstances, and for new and revised data as it comes to hand.

An example of a small but practical network is given in detail in Chapter 8.

Section 3 Selection and management of the project planning team

Definition of the project and objectives

As with any enterprise it is essential at the outset to be quite clear on what the project and its objectives are before planning starts.

At some level in the management organization depending on the type of work and size of the organization it will be the responsibility of some body or individual to approve and initiate the project. It is his responsibility also to define clearly what the project is, what are the start and end points, what external factors may affect its execution such as intermediate or end target dates, requirements of other departments and projects which may give rise to limitations in the availability of men, machines, materials and money, and so on. Without a knowledge of these constraints and objectives it will be difficult for those charged with the execution of the project to plan with any degree of assurance.

These factors will be ascertained and decided in discussion with all who may influence or be affected by the project. This is not to say these decisions may not be changed in the light of changing circumstances but let them be stated clearly and agreed at the outset by all concerned.

Selection of project planning team for large projects

The initiator of a large project must appoint a Project Planning Officer to whom will be delegated the responsibility for planning and co-ordinating all the activities and people concerned in the project. It is of the utmost importance to see that the Project Planning Officer has delegated to him sufficient authority to enable him to discharge these responsibilities. The Project Planning Officer must have similar status and experience to the Project Manager himself in order that they may work together to the maximum effect and derive the maximum advantage from the planning system which is to be set up on the basis of network analysis. It must remain clear however that the Planning Officer is only acting on behalf of the Project Manager who must at all times retain over-all responsibility for the project.

The Planning Officer must then proceed to select a team of men who will actually draw the arrow diagrams and set up the planning system under his guidance. Each department or function concerned in the project must be represented on the team by a member of that department armed with authority, and relieved of other responsibilities, to the extent necessary to enable him to contribute to the planning effort all the necessary details of his department's stake in the project. This means in effect that the arrow diagrams and planning system will be set up by people acting on behalf of the departmental personnel responsible for carrying out the activities in the network. In this way the network planning system will be an integrated part of the management process. Co-ordination will rest with the Planning Officer acting on behalf of the Project Manager.

As with any planning system the sooner the plan is set up the greater the advantage. If the plan can be set up and analysed before work actually starts on the project, so much the better.

A couple of examples may serve to illustrate these points. In a project involving the reconstruction and electrification of a stretch of railway line, the project planning team would consist of representatives of the Civil Engineering, Signal and Telecommunications, Electrical and Traffic Departments, the last mentioned being required mainly to advise on minimizing the effects of reconstruction on existing traffic flow.

Where a decision has been made to launch a new engineering product on to the market it will be necessary to have represented on the planning team the Marketing and Sales Department, Research and Development Department, Engineering Department (for making jigs and tools). Production Department and Warehousing and Despatch Department.

In either case all those departments which play a major role in the projects are represented on the planning team. Having had the plan prepared therefore by personnel acting on behalf of those responsible for the project difficulties in gaining acceptance of the plan are less likely to arise.

Personnel for planning small projects

For small projects confined, say, to a single department in which one manager or foreman has direct control of all the people involved in the project—a number of different tradesmen, for instance —it is clearly unnecessary to set up a planning team. The manager and foreman would in the past have had to plan the project by themselves using traditional planning methods. Now they can use Network Planning Techniques as part of their daily routine to improve planning and scheduling of many of the projects in their own department. An obvious example is a project involving the servicing or overhaul of expensive machines or plant when time out of commission represents a heavy loss of revenue.

Whether the project is large or small Network Planning Techniques are powerful tools in the day-to-day management process in project work.

Training

Formal classroom training in the principles and uses of Network Planning Techniques need not exceed one week at the most. The principles are simple enough. The difficulties of applying them in practice can be learned only by experience and the best means of getting this experience is by working with a team of network planners under the supervision of someone who already has experience.

The British Institute of Management can provide information about training courses available in various parts of the country.

Deciding strategy—the 'master plan'

The first job of the Project Planning Team is to obtain from Management agreement on strategy. Whether the project is large or small there may be several possible over-all methods of tackling it and one of these must be chosen by the Project Manager as apparently the best strategy to employ in the light of his previous experience of similar projects.

At this stage, only the very broadest considerations are relevant. For instance in the electrification of a length of railway line it would be necessary to consider at this time how to split the length of line into stages, the order in which the stages must be carried out, and so on. In the case of launching a new product on to the market it would be necessary to consider how to organize publicity and at which factory to manufacture the product. In the case of overhauling an expensive machine or plant, the availability of special tools or equipment would be considered.

In each case the agreed strategy should be represented by a relatively simple, broad-scale network diagram. This diagram will then act as a 'master plan' for the project.

Expansion to detailed network diagram

From the 'master plan' network it will then be necessary to produce more detailed network diagrams. For a small project the whole project would probably be expanded into a more detailed diagram. For a large project it may be possible to arrange for individual members of the planning team to prepare detailed networks for various sections of the project more or less independently and link these together afterwards. The 'master plan' network will indicate which sections should receive priority for detailed study in this way. As always the degree of detail and the choice and size of the activities must be related to the needs of the ultimate user of the results of the network analysis.

Such diagrams may have anything from less than one hundred up to several thousands of activities. The more complex the diagram the fewer the number of events in relation to the number of activities. There is however no relationship between the size of the project and the size of the arrow diagram. The arrow diagram will be prepared in detail only sufficient to give the degree of planning control sought at any particular level of management.

Time required to prepare networks

It is impossible to give any very precise guidance on how long it will take to prepare the network diagram required to control a project.

With a suitably constituted team suitably trained in the technique and experienced in its application, it should be possible to complete a 1,000-activity network of moderate complexity in less than one month. Larger projects would require larger teams of people to prepare the networks so that in fact very large projects indeed, representing several millions of pounds worth of capital investment for instance, can have networks prepared in under three months.

Conclusion

It may appear from the trivial examples used in this chapter that the preparation of a network diagram is a simple matter. This is not so. The preparation of a network diagram forces the people concerned to take a new look at the activities in the project in a disciplined logical way, and this inevitably brings to light many problems which would not otherwise be foreseen. These problems would however have to be faced and solved eventually. The preparation of an arrow diagram has the merit of drawing attention to them before they reach crisis proportions.

The technique has been devised for, and is most powerful in dealing with, very complex projects. In preparing an arrow diagram these complexities are committed to paper. There is less need therefore for any individual to try to remember them all, the project can be planned in much greater detail, and the arrow diagram acts as an excellent means of communicating them to all concerned. Everyone is therefore working to a common plan.

It is important to emphasize again in concluding this chapter that the preparation of the network diagram is the cornerstone of the whole technique. The logic built into the network must be a true and clear representation of the interrelationships between the constituent activities in the project in accordance with the method or strategy by which the project is to be tackled. The network must therefore be prepared by personnel completely familiar with the technologies involved, and with sufficient delegated authority to enable them to have the results of their planning analysis accepted at the site of operations. Only in this way can the success of the use of the techniques be reasonably assured.

Four Preparation of estimates

Introduction

No planning system can be set up on a quantitative basis without numerical data relating to the activities in the project. Besides describing the logical sequences of activities in the project by numbering the events in the network, estimates of activity durations are required before analysis of the network is possible.

Estimates of activity costs, and requirements of manpower and other resources for each activity, may also be required in order to provide a sound basis for a comprehensive planning system based on the minimum cost and the most efficient use of resources possible.

It may not be essential, however, to provide all these estimates at the outset or even at all. Clearly the type of information required depends on the objects of the network planning system and the degree of control required. If the objective is only to search for the shortest time schedule for the project, estimates of activity durations alone may be sufficient. If the availability of some of the resources such as money or manpower is likely to be a problem, estimates of costs and manpower requirements for each activity may also have to be prepared.

Methods of estimation is a subject in itself and it is not intended to develop it in any great detail here. The purpose of this chapter is merely to make a few general observations concerning what is required in the way of estimates of times, costs and resource requirements in particular network planning situations rather than to discuss in detail methods of producing these estimates. The reader wishing to make a detailed study of Work Study methods of estimation is referred to such standard works as *Motion and Time Study* by Ralph M. Barnes, *Motion and Time Study—Principles and Practice* by M. E. Mundel, or *Introduction to Work Study*, International Labour Office, Geneva, 1962.

Estimates required

An estimate of the duration of each activity in the network must be prepared before an analysis of the time implications of the network can be made. For instance it will clearly be impossible to calculate the over-all time the project will take without first estimating the durations of all the activities.

There is no special way of preparing an estimate of duration in network planning. It is an estimate in the traditionally accepted sense and must be prepared on the assumption that 'normal' conditions will obtain, that the method 'normally' accepted as the most economic will be used, and that the 'normal' quantities of resources will be available.

The method of preparation should of course be analytical and systematic, and not mere guesswork. Breaking the activity down into its constituent elements and aggregating the elemental times supplied from Work Study sources, provides a completely satisfactory means of producing these estimates in many cases. If, however, elemental times are not available from Work Study sources they can be estimated.

For a fairly large activity it may be helpful to expand the activity into a detailed individual network as an aid to preparing the estimate of the activity duration. One of the advantages of using Network Planning Techniques is that a more reliable forecast of over-all project time can be made. This property of network techniques can obviously apply in preparing an estimate of the duration of an activity which can itself be represented by a complete arrow diagram on a finer scale of detail.

However the estimate of duration is prepared the result will imply the use of a certain amount of manpower, equipment and other resources. If therefore the estimates of durations are prepared systematically, estimates of resource requirements should be an automatic by-product, and will therefore be available for use if required for scheduling and resource allocation procedures.

The preparation of cost information relating to each activity is optional. It depends on the nature of the project, the special objectives of the planning system, and the magnitude of the direct cost of the project in relation to the indirect costs (overheads and loss of revenue). Clearly if indirect project costs are so great as to dwarf direct costs no analysis of direct costs requiring the preparation of individual activity costs is likely to be of great value. Concentration on reducing over-all project time in order to minimize indirect costs will produce the greatest return for effort. If on the other hand direct and indirect project costs are of the same order of magnitude, it may pay to prepare individual activity costs in order that an analysis of over-all project costs may be made with a view to optimization. (See also Critical Path Method, in Appendix 1.)

Estimators

It is essential that the estimate for an activity should be prepared by personnel responsible for, or acting on behalf of those responsible for, the execution of the activity. They are not only the best qualified to prepare the estimate but are less likely to commit themselves to a wildly unrealistic estimate. This does not mean of course that the estimate may not be inaccurate for various reasons but the system of holding the executive responsible for preparing estimates for his own activities militates against excessive inaccuracies, and makes it more likely that he will stand by them.

Accuracy of estimates

Although it is true to say in general that estimates should be as accurate as possible the need to strike a balance between higher accuracy on the one hand and the cost of achieving it on the other, must be borne in mind.

An estimate is by definition a forecast approximation and must be accepted as such. In preparing an estimate there comes a point beyond which it is no longer economic to continue trying to improve accuracy. This is particularly relevant in the context of Network Planning Techniques. The real necessity is to produce a first estimate of each activity as quickly as possible to enable the network diagram to be analysed quantitatively. If this first estimate has to be produced at some sacrifice of accuracy this will not necessarily be a serious disadvantage. The subsequent analysis of the network will identify those activities which are critical in that they govern the over-all length of the project and for which estimates should therefore be as accurate as possible. Special attention should then be given to revising these estimates rather than dissipating the efforts of estimators on devising over-accurate estimates for non-critical activities. The accuracy of a two-week estimate of duration is likely to be less important if the activity has a float of say twenty weeks. This is not to say, however, that obvious inaccuracies should not be eliminated—far from it.

In preparing estimates of costs and other resource requirements there is no distinction between the degrees of accuracy required for different activities. Whether or not the activity is critical its cost will be part of the total project cost and its requirement of resources will be part of the total requirement. These estimates should therefore be as accurate as possible in any case. Indeed the predicted total cost and resource requirements for the project as a whole will be only as accurate as the original estimates for each activity.

Activities with a large measure of uncertainty

The preparation of estimates for activities in which there is a large measure of uncertainty, as for instance in research and development work, presents a special problem. Estimates may be little more than inspired guesses in some cases and nothing better is possible.

There is however one particular network planning system known as PERT (Programme Evaluation and Review Technique) which was originally devised to cope with the uncertainties in the development of the Polaris missile. This system requires the preparation of three estimates of duration, the optimistic, the pessimistic and the most likely. Each of these three estimates has associated with it a probability that the activity will actually be performed in that time. The three estimates and their probabilities are combined to calculate an 'expected' duration for the activity.

The PERT system is discussed in more detail in Appendix 1 at the end of this book.

In the straightforward project in practice however it is usually possible to provide estimates of duration for most of the activities to an acceptable level of accuracy based on previous experience. A network may however contain a small number of activities whose durations cannot be meaningfully estimated, as for example : 'obtain planning permission'. When this occurs it is possible to adopt the practice of giving these activities zero duration in the first instance. The preceding and succeeding event times resulting from the analysis of the network (see Chapter 5) will then indicate at least how much time is available for the completion of the activity in question. A target date for completion of the activity is thus introduced which if not met could delay the completion of the project.

Time units

The time units in which the estimates of duration are expressed may be hours, days, weeks or months as appropriate.

It is often convenient to assume a five-day working week keeping weekends for speeding-up if required. Since one-tenth of a week is then half a day it is possible to express durations in decimal time units.

All the durations in one diagram must of course be expressed in the same time units (days, weeks, etc.).

Entry of estimates on the arrow diagram

It is customary to write the estimate of duration of each activity against the appropriate arrow in the network diagram, and it is sometimes enclosed in a box thus (time unit = 1 week):

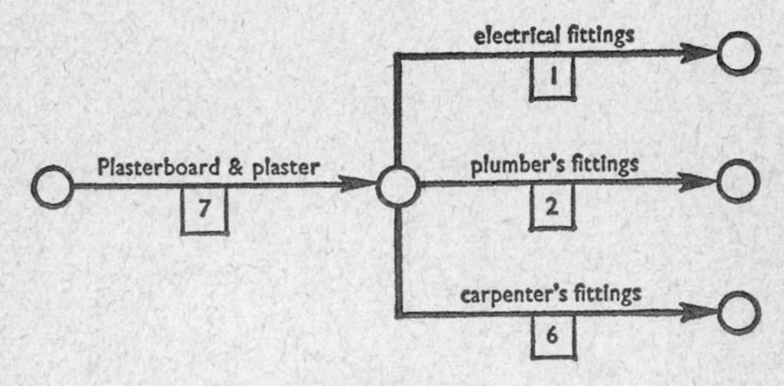

Figure 4.1

Dotted arrows (dummies) do not have durations written against them as it is understood that the duration is zero.

Estimates of resource requirements and costs for each activity may also be written against the appropriate arrow in the network if desired. It is certainly an advantage in practice to have all the necessary information available for quick reference on the network diagram itself. The main limitation on adding information to the network is of course space, but if the space is available all the estimates may be written on the diagram provided they are distinguished from each other and from other figures which may later appear on the diagram, either by different colours or by enclosing them in differently shaped boxes.

When the estimates have been written on the network diagram the diagram is then ready for analysis. This is the subject of Chapter 5.

Five Analysis of the network diagram—the critical path

Introduction

A major step in setting up a project network planning system has been accomplished with the completion of the network diagram and the preparation of the necessary estimates of activity durations, costs and resource requirements.

In order to make use of the system the network diagram must now be analysed.

The analysis of the network diagram will provide data which will enable the planners and managers of the project to set up a precise and efficient schedule for the project and monitor and control its progress.

The analysis of networks is relatively simple in principle but involves the performance of many arithmetical calculations. The number of calculations depends on the size of the network and the amount and type of information required from the analysis. The calculations may be carried out manually or by computer according to circumstances.

The first section of this chapter is devoted to an explanation of the mechanics of analysing networks manually and the data this generates. The second section gives an introduction to the use of computers for carrying out this analysis and the third section discusses the merits of using these different methods of analysis in different sets of circumstances.

Section 1 Direct analysis of the network diagram

1 Calculation of event times

The first step in the analysis of a network diagram is to calculate the earliest and latest times at which each event can occur.

The earliest time at which an event can occur is determined by the longest (time) path from the start event up to the event in question, because there must be time to complete the longest chain of activities before that event can be reached. The latest time an event can occur, if the end of the project is not to be delayed, is determined by the longest (time) path from the terminal event back to the event in question, because time must be allowed to complete the longest chain of activities from that event to the terminal event.

EARLIEST EVENT TIME (ET)

The earliest time of the start event can be taken as zero. The earliest possible time for each succeeding event along each path can be calculated by successively adding intervening activity durations. Where there is more than one possibility for the earliest time of an event representing more than one path leading to that event, the largest total time (representing the longest path) is the earliest event time. Thus:

$$\text{earliest event time} = \text{maximum of} \left\{ \begin{array}{c} \text{earliest time of} \\ \text{preceding event} \end{array} + \begin{array}{c} \text{intervening} \\ \text{activity} \\ \text{duration} \end{array} \right\}$$

An example will serve to illustrate the method of carrying out these calculations. If event numbers and durations (days) are added to the network obtained in the answer to Exercise 5 in Chapter 3, and the event circles are extended thus: to provide two boxes, the upper for the earliest event time and the lower for the latest event time, the network will look like this:

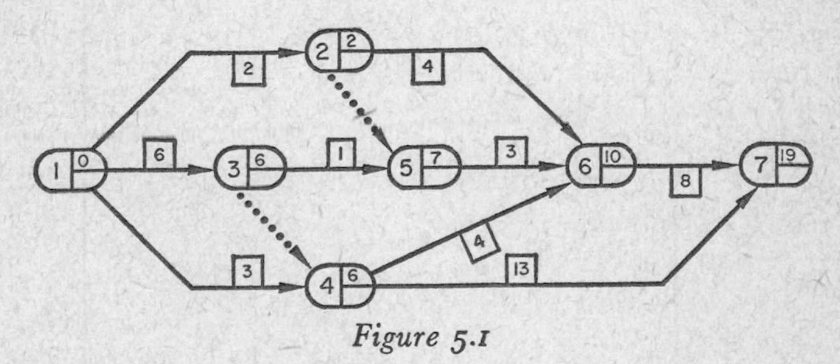

Figure 5.1

The earliest time of event 1 is taken as zero and a '0' is therefore placed in the upper box attached to event 1. There is only one path from event 1 to event 2 with an intervening activity of duration 2. The earliest possible time for event 2 is therefore $0+2 = 2$. Similarly the earliest time for event 3 is $0+6 = 6$.

There are two paths to event 4, one from event 3 and one from event 1. The earliest time for event 4 is therefore the larger of

6+0 and 0+3, namely 6, representing the longest path from the start event to event 4.

The earliest time for event 5 is the larger of 2+0 and 6+1, namely 7. The largest of the three possibilities for event 6 is 10 and for event 7 there are two possibilities, the larger giving an earliest event time of 19.

Event 7 is the terminal event in this simple example and its earliest time, 19, is the minimum length of the project under the conditions implied by the network diagram.

LATEST EVENT TIME (LT)

If the end of the project is not to be delayed beyond the earliest completion time, the latest time for the terminal event must be the same as its earliest time. The latest possible time for each preceding event along each path backward from the terminal event can be calculated by successively deducting intervening activity durations. Where there is more than one possibility for the latest time of an event, representing more than one path back to that event, the smallest possibility (representing the longest path to the end event) is the latest event time. Thus:

$$\text{latest event time} = \text{minimum of} \left\{ \text{latest time of succeeding event} - \text{intervening activity duration} \right\}$$

In the example in Figure 5.1, the latest time for terminal event 7 is the same as the earliest time, namely 19, and is shown in the lower box attached to event 7.

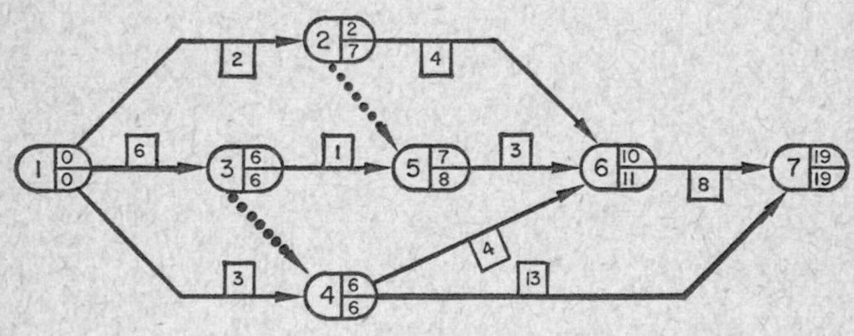

Figure 5.2

There is only one path from event 7 back to event 6 with an intervening activity of duration 8. If the project is to be completed in 19 days the latest possible time for event 6 is therefore 19 − 8 = 11.

Similarly for event 5 the latest time is 11 − 3 = 8.

There are two paths back to event 4, one from event 6 and one from event 7. The latest time for event 4 is therefore the smaller of 11–4 and 19–13, namely 6, representing the longer time path from event 4 to the end of the project.

The latest time for event 3 is the smaller of 6–0 and 8–1 namely 6. Similarly for event 2 there are two possibilities the smaller giving a latest event time of 7. The smallest of the three possibilities for event 1 is 0.

These results are summarized in the following table:

Event No.	1	2	3	4	5	6	7
Earliest Time	0	2	6	6	7	10	19
Latest Time	0	7	6	6	8	11	19

Figure 5.3

CRITICAL EVENTS—EVENT SLACK

The earliest and latest possible times of occurrence of all events have now been calculated. Those events for which the earliest and latest times are the same are critical events in the sense that there is no flexibility, i.e. zero slack in the time at which each can occur. In this example events 1, 3, 4 and 7 are critical.

The slack associated with each event is simply the difference between the earliest and latest time for the event. Thus:

$$\text{event slack} = \begin{matrix}\text{latest time}\\ \text{of event}\end{matrix} - \begin{matrix}\text{earliest time}\\ \text{of event}\end{matrix}$$

The slack for each event can be recorded in a further row at the foot of the table of event times (Figure 5.4).

Event No.	1	2	3	4	5	6	7
Earliest Time	0	2	6	6	7	10	19
Latest Time	0	7	6	6	8	11	19
Event Slack	0	5	0	0	1	1	0

Figure 5.4

In all these calculations it will be seen that the dummies are treated in all respects as activities of zero duration.

CRITICAL ACTIVITIES—CRITICAL PATH

An activity is critical if there is no flexibility in the time at which it may be carried out. That is to say an activity is critical if:

(*a*) it is immediately preceded and followed by a critical event,
and (*b*) the difference between the times of the events immediately preceding and immediately succeeding the activity is equal to the activity duration.

Thus in the example, activities 1–3, 3–4, and 4–7 are critical but because the difference between the times of events 1 and 4 is 6–0 = 6 and the duration of activity 1–4 is 3, activity 1–4 is not critical.

The chain of critical activities from the start event to the terminal event is the longest (time) path through the network. This chain or sequence of critical activities governs the length of the project as a whole and is therefore called the Critical Path. The project can be shortened only by shortening the critical path, and effort applied to shortening non-critical activities will have no effect on the length of the project.

Event times are of great value because instead of having only one or perhaps a few target dates to work to in the project, the event times provide target dates for each and every activity in the network. This makes control much easier and helps to ensure that forecast project completion dates are achieved.

2 Calculation of activity times

In many types of project it is easier to think in terms of activities rather than events. The project manager and foreman want to know at what time activities can be started and finished. These times can be taken from the arrow diagram and tabulated.

EARLIEST STARTING TIME (EST)

An activity clearly cannot start until the event from which its arrow emanates has been reached. The Earliest Starting Time for an activity is therefore the earliest time of the preceding event.

Referring again to the demonstration network in Figure 5.2 the EST of activity 2–6 would appear to be the earliest time of event 2 namely day 2. It is worth making the point here however that this actually means that activity 2–6 can start at the earliest, at the *end* of day 2 which means in effect a start on the *morning* of day 3. This is a consequence of giving the first event the time zero, so that starting activities have EST's of zero, i.e. they can start on the morning of day 1.

For the purpose of the calculations however:

EST of activity 2–6 = earliest time of preceding event = 2.

EARLIEST FINISHING TIME (EFT)

The Earliest Finishing Time for an activity is clearly the duration further on from its earliest starting time. Thus for activity 2–6, the earliest finishing time = EST + duration = 2 + 4 = 6. This means that activity 2–6 can finish at the earliest at the *end* of day 6.

LATEST FINISHING TIME (LFT)

An activity must finish before its terminating event can be reached. The Latest Finishing Time (LFT) must therefore be the latest time of the terminating event. Thus the latest finishing time of activity 2–6 is the latest time of event 6 namely day 11. Again this means the *end* of day 11.

LATEST STARTING TIME (LST)

The Latest Starting Time of an activity is clearly the duration ahead of its latest finishing time. Thus for activity 2–6 the latest starting time is LFT–duration = 11–4 = 7. Again, in practice this means that activity 2–6 can be started at the latest after the end of day 7, that is on the morning of day 8.

For each activity there are therefore four times to be calculated:

(i) Earliest Starting Time (EST) = earliest time of preceding event.
(ii) Earliest Finishing Time (EFT) = EST + activity duration.
(iii) Latest Finishing Time (LFT) = latest time of following event.
(iv) Latest Starting Time (LST) = LFT – activity duration.

It is understood that the activity times are at the end of the day (or other time unit) so calculated, and for starting times this is normally equivalent to the morning of the next day.

These four times for all the activities in the network in Figure 5.2 can now be calculated, and tabulated thus:

Activity	Duration	EST	EFT	LST	LFT
1 – 2	2	0	2	5	7
1 – 3	6	0	6	0	6
1 – 4	3	0	3	3	6
2 – 5	0	2	2	8	8
2 – 6	4	2	6	7	11
3 – 4	0	6	6	6	6
3 – 5	1	6	7	7	8
4 – 6	4	6	10	7	11
4 – 7	13	6	19	6	19
5 – 6	3	7	10	8	11
6 – 7	8	10	18	11	19

Figure 5.5

As all activities must be completed before the project is complete, the largest EFT defines the minimum time for the project if carried out by the method implied in the arrow diagram.

Again dummies are treated as activities of zero duration.

Activity float

(A) TOTAL FLOAT (TF)

The Total Float associated with an activity is the total available flexibility in the time within which it must be carried out if it is not to affect the over-all time for the project. In order to avoid affecting the time for the project each activity must clearly be carried out between its EST and LFT. The total flexibility for each activity is therefore the interval between its EST and LFT minus the activity duration, i.e.

$$
\begin{aligned}
TF &= LFT - EST - D \\
&= LFT - (EST + D) \\
&= LFT - EFT \\
\text{or } TF &= LFT - EST - D \\
&= LST - EST
\end{aligned}
$$

That is to say the Total Float for an activity is the difference between its earliest and latest starting times or finishing times. This is illustrated for activity 2–6 in Figure 5.6.

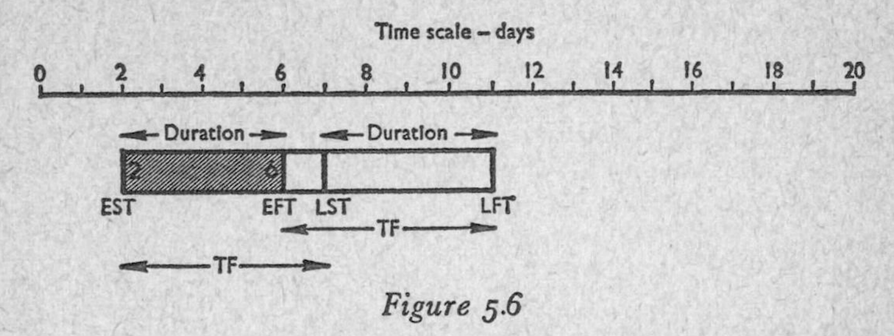

Figure 5.6

Another way of looking at the calculation of Total Float is direct from the network.

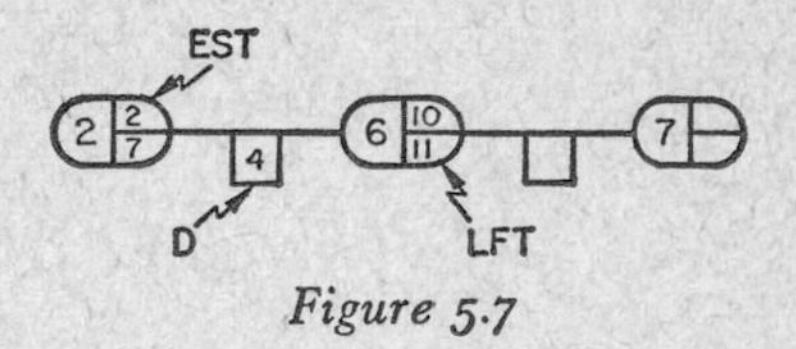

Figure 5.7

It has been noted that

$$TF = LFT - (EST + D)$$

so for activity 2–6

TF equals (EST + D) away from LFT
i.e. 2 + 4 away from 11
i.e. TF = 5

A column showing Total Float for each activity may be added to the right of the table of activity times as shown in Figure 5.10.

The critical activities are clearly those with zero Total Float.

(B) FREE FLOAT (FF)

The Free Float associated with an activity is the flexibility in the time it may be carried out if all immediately succeeding activities are carried out at their EST's, i.e., the time an activity can be delayed without delaying any succeeding activity. If the succeeding activities start at their EST's, then the activity under consideration has flexibility only to the extent that it must be carried out in the time interval between its own EST and the EST of the succeeding activities. The Free Float for an activity is therefore this time interval minus the activity duration, i.e.

$$\begin{aligned} \text{FF} &= \text{EST}_{\text{succ.}} - \text{EST} - \text{D} \\ &= \text{EST}_{\text{succ.}} - (\text{EST} + \text{D}) \\ &= \text{EST}_{\text{succ.}} - \text{EFT} \end{aligned}$$

That is to say the Free Float for an activity is the difference between its EFT and the EST of the succeeding activities. This is illustrated for activity 2–6 in Figure 5.8.

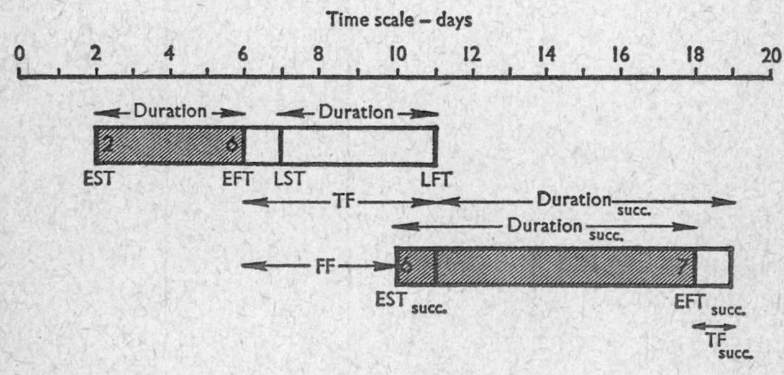

Figure 5.8

Another way of looking at the calculation of Free Float is direct from the network.

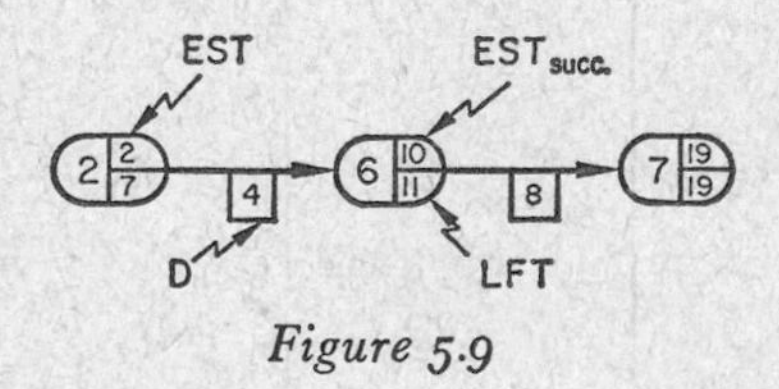

Figure 5.9

It has been noted that

$$\text{FF} = \text{EST}_{\text{succ.}} - (\text{EST} + \text{D})$$

So for activity 2–6 :

FF equals (EST + D) away from $\text{EST}_{\text{succ.}}$.

i.e. 2 + 4 away from 10
i.e. FF = 4

Free Floats can be shown in a further column to the right of the table thus:

Activity	Duration	EST	EFT	LST	LFT	TF	FF
1 – 2	2	0	2	5	7	5	0
1 – 3	6	0	6	0	6	0	0
1 – 4	3	0	3	3	6	3	3
2 – 5	0	2	2	8	8	6	5
2 – 6	4	2	6	7	11	5	4
3 – 4	0	6	6	6	6	0	0
3 – 5	1	6	7	7	8	1	0
4 – 6	4	6	10	7	11	1	0
4 – 7	13	6	19	6	19	0	0
5 – 6	3	7	10	8	11	1	0
6 – 7	8	10	18	11	19	1	1

Figure 5.10

Clearly the Free Float for an activity can never be greater than its Total Float. Activity 2–6 can be delayed from finishing at day 6 to finishing at day 10 (Free Float = 4) without affecting any other activity. If, however, full advantage is taken of the Total Float of activity 2–6 to delay its completion to day 11 this robs the succeeding activity 6–7 of its Total Float of 1 day. In other words Total Float is associated with sequences rather than individual activities and when advantage is taken of the Total Float associated with one activity this will reduce or wipe out the total Total Float associated with subsequent activities in the sequence.

The activity times and floats for all the activities in the network in Figure 5.2 are shown against a time scale in the following diagram:

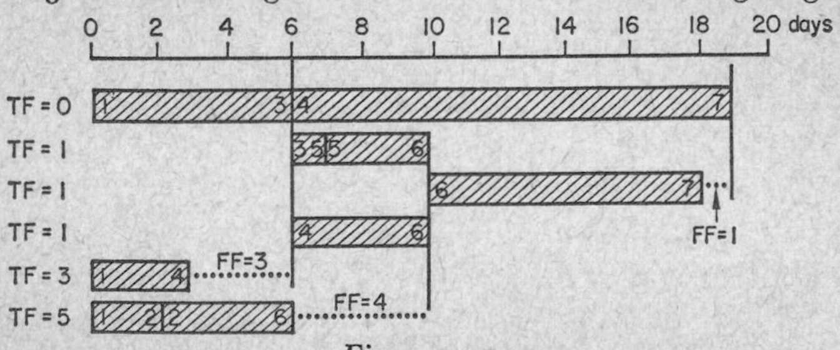

Figure 5.11

This type of bar chart presentation may help to clarify the meaning and significance of the times and floats associated with individual activities and how these depend on the interrelationships between the activities in the network. In a modified form it can also be used as a means of constructing the network against a time scale, with vertical lines to indicate linkages between activities. Clearly however this has limitations because of the sheer physical difficulties involved in laying it out on paper.

There is now in the table in Figure 5.10 all the information about the activity times which is required for preparing a schedule for the project. The floats show to what extent the timing of the activities can be altered either accidentally or by design without delaying the completion of the project. The project planners therefore know from the results of the network analysis the extent to which they may vary the timing of any activity when preparing the schedule for the project and this is particularly helpful when considering smoothing the demand for resources. Scheduling and resource levelling is the subject of the next chapter, and a schedule for a practical project is developed in Chapter 8.

More about the critical path

A critical activity (see page 36) is one which is contained between a pair of critical events and whose duration is equal to the difference between the times of the two events. This means that if the project is to be completed according to the network plan there is no choice in the timing of the critical activities—they have no float (see page 39).

The sequence of critical activities, i.e. the critical path, forms the longest chain of activities from the start to the end event and therefore governs the over-all time of the project. Along part of all of its length the critical path may split into two or more parallel paths and there is no reason why dummy activities should not form part of the critical path.

The identification of the critical path is clearly a very important first step in the planning and management of a project. It will immediately direct planning and management attention to those activities which govern the completion date of the project.

It should not be supposed however that activities can be subdivided into two groups, those that are critical and those that are non-critical. The activities in any project can be ranked in ascending order of total float, that is in ascending order of importance to the early completion of the project, and these floats will range from zero upwards. A prudent manager will concentrate his attention not only on the zero float critical activities but also on the small float 'near critical' activities. In fact he will usually find that these will be relatively few, numbering not more than about 5 per cent of all the activities in the network.

First analysis of network

In practice the first attempt at analysing a network will usually throw up not a critical path but a series of anomalies which will have to be rectified before a proper analysis can be completed. These anomalies may include duplication in numbering of events, or of activities, or a 'dangle' i.e. an arrow whose head has been left in the air instead of terminating in an event, making it appear that the project has more than one end point. Logical 'loops' (see page 14) may also come to light during the analysis of the network.

When these anomalies have been rectified by referring back to the network itself it will be possible to complete the analysis and identify the apparently critical activities. The critical path first thrown up may cause some surprise and consternation, because of its length and content; the project cannot surely take this long? these are surely not the critical activities? It has the merit however of now concentrating the attention on those areas of the project where refinement of logic and durations is most important to the preparation of a successful plan for the project.

Negative float

It is often the case in practice that external factors may require fixed target dates to be imposed on intermediate events in the network as well as on the start and final events.

It may be, for example, that the new office premises in a project have to be ready before the termination of the lease on the existing office premises whilst there is no corresponding need for the early completion of the factory premises. Or perhaps a series of drawings has to be completed by a particular date to enable an application for planning permission to be considered by a particular council meeting. These management objectives imply fixing target dates for particular events in the network.

Such target dates can now be substituted for the latest times for the events in question. If the target date for any event is earlier than the calculated latest time for that event then the network as it stands may not meet the target date. Some activities will have negative float and it is these activities which must be shortened if negative float is to be eliminated and target dates are to be met.

Returning to Figure 5.2 for example, if event 5 had to be reached not later than the 7th day activity 3–5 would, in effect, be critical although activity 5–6 would be unaffected. If however event 5 had to be reached by the 6th day the analysis of the network would look like this :

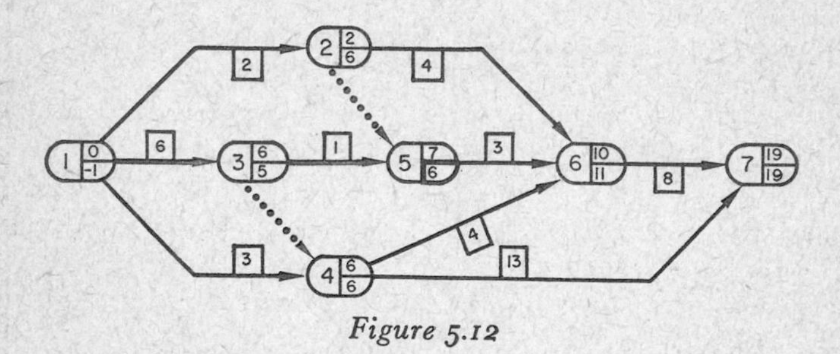

Figure 5.12

Activities 1–3 and 3–5 would now have negative float (TF = –1). The duration of one of these activities (obviously activity 1–3 in this case) would have to be reduced by one day in order to meet the target date imposed on event 5.

It may be noted in passing that the concept of a straightforward critical path running from the start event to the end event may be an over-simplification of what can happen in practice if target dates are imposed on intermediate events. Such target dates can give rise to negative float which has to be eliminated before the network has any meaning.

Refining the network to meet management objectives

In refining the network in practice to meet management's objectives and targets, and to eliminate negative float, it is clearly necessary to concentrate effort on the critical and near-critical activities.

(I) RE-EXAMINING DURATIONS

One of the steps towards refining the network will be to confirm by whatever means are necessary that the estimates of duration for the critical activities are accurate. This may involve checking that those who provided the estimates did not build into them unreasonable allowance for contingencies. Or on the other hand it may involve taking a single or small group of activities from the network and 'exploding' it separately into a more detailed network to ensure that the original estimate is achievable in practice. For example the construction of a whole building may be represented in the original network by a single arrow of 20 weeks. A separate detailed arrow diagram could be drawn for the building alone in order to confirm the duration.

(II) RE-EXAMINING METHODS AND PROCEDURES

Perhaps of more importance in refining the network is a re-examination of the methods of working implied by the logical sequences of critical and near critical activities in the network. It is at this stage that the best methods of operation for the critical activities will be sought by a careful examination of possible alternatives not excluding, if only by inference at this stage, a change in and more precise statement of the resources (men, machines, etc.) which would be required. The cheapest method for each activity may not always be chosen if a costing exercise can demonstrate that the higher cost of an alternative method will be justified by savings resulting from earlier completion of the project, as for example prefabrication instead of construction *in situ*. Any change in methods or procedures would of course involve a change in the logical relationships between arrows in the network which would therefore have to be revised. The re-examination of critical delivery programmes from external suppliers is a particular and usually highly fruitful example of the type of refinement which can be achieved.

Also at this stage every effort would be made to improve the logical representation of the interrelationships involving the critical activities, including the replacement of single arrows or groups of arrows in the original network with overlapping arrows as described on pages 10–13. Some of the activities for example following the construction of the building mentioned in the previous paragraph may be started when the building is only partly completed. Again, these detailed studies would normally be confined to the critical activities.

(III) TIME REDUCTION AT ADDED COST

It may also be at this stage that some costing exercises will be undertaken. The duration of most activities can usually be reduced at some additional cost. If the cost savings, direct and indirect, from advancing the project completion date are greater than the cost of accelerating a particular activity, as for example by prefabrication instead of construction *in situ*, then clearly the extra cost of doing so will be worth while. (This principle, extended and formalized, is the basis of the CPM system of network planning briefly described in Appendix 1.)

Evolution of network plan

By a process of alternately refining and re-analysing the network, during which the apparent critical path will shift from one sequence to another, a network will ultimately be evolved which will meet all the objectives and which, because it *looks* right, will no longer cause surprise and consternation. For a large project (say 1,000 activities) the network may have been analysed perhaps five or six times before this point is reached and will now represent a practical network for the project taking management's objectives into account. It may not necessarily be the best method throughout but so long as it meets management's objectives it can be presented to the Project Manager as a practical network. It is for him to say whether he would wish more time and effort to be spent seeking further improvements, time and effort which at this stage are likely to prove substantial for the achievement of a very limited improvement in the network.

The network at this stage still takes no account of any limitations in the availability of resources which may exist in the practical situation. Making the necessary adjustments to the network to take account of these limitations, and to improve utilization of resources by improved scheduling, is the subject of Chapter 6.

Exercises

7 If the activities in the network in Exercise 3 are given durations as follows, calculate the duration of the project and identify the critical events and activities.

Activity		Duration
	Event Nos.	days
N	1 – 2	6
A	2 – 3	3
C	2 – 4	5
B	2 – 5	1
D	3 – 9	8
L	4 – 9	6
J	5 – 6	1

Activity		Duration
	Event Nos.	days
F	5 – 7	4
H	5 – 8	10
K	6 – 8	7
G	7 – 8	2
E	8 – 9	3
P	9 – 10	2

Tabulate the EST, EFT, LST, LFT, TF and FF for all the activities.

Exercise

8 If the activities in the network in Exercise 4 are given durations as follows, calculate the project duration and identify the critical events and activities :

Activity		Duration days
B	1 – 2	4
A	1 – 3	3
C	3 – 4	2
D	3 – 5	6
E	3 – 10	2
F	4 – 8	1

Activity		Duration days
K	5 – 11	4
G	6 – 7	3
H	7 – 9	5
L	8 – 11	3
I	9 – 11	4
J	10 – 11	2

Tabulate the EST, EFT, LST, LFT, TF and FF for all the activities.

Section 2 **Calculations by computer**

The first section of this chapter has dealt with the mechanics of carrying out the analysis of networks by manual methods and these methods are used in many projects in practice. They do however involve a great deal of repetitive arithmetic and circumstances can arise when the sheer volume of work becomes uneconomic, if not impossible. In these circumstances it is possible to consider using a computer to carry out the calculations. This section gives first a general description of the nature of the computer and the requirements for using it, followed by several paragraphs on the use of computers specifically in the field of network analysis. If the reader is already familiar with the general nature of the computer he may wish to omit the next paragraph and proceed directly to Programmes for network analysis on page 50.

The nature of the computer

Calculations of the kind required in network analysis can be carried out on a digital computer of which there are now many different sizes and makes.

It is not an over-simplification to say that a computer is merely a high-speed desk calculating machine with the exception that a computer can be given a programme of instructions which will enable it to carry out long series of calculations, choosing between alternative courses of action at particular points according to predefined criteria, without awaiting further instructions from the operator. This ability to choose between alternatives adds enormously to the power and flexibility of computers.

In the case of an operator working with a desk calculating machine the operator's brain, working through her fingers, provides the flow of decisions and instructions as to what calculations are to be carried out on the machine. With a computer it is necessary to provide, in advance, a programme of instructions as to what calculations are to be carried out in the machine. If the programme is followed into the computer by numerical data, the machine will carry out the calculations specified by the programme, using the data supplied, and refer back to the programme for further instructions when necessary.

Programmes of instructions for difficult computational operations can be immensely long and complex when written in terms of very large numbers of combinations of the four basic arithmetical instructions: add, subtract, multiply and divide, along with the additional instruction: 'choose between alternatives'. Even using the 'subroutines' now available for most computers the preparation of a computer programme can be a long and exacting task.

If an operator were required to carry out equally difficult computations with a desk calculating machine she would have to be similarly instructed. But because the operator's speed is limited, and because she may have to make a note of the results of intermediate calculations on a note-pad for later reference, long and complex operations can for her be very time-consuming indeed. A computer can carry out the same operations at immense speed and being provided with internal storage space for holding the results of intermediate calculations does not have to stop before all the instructions in the programme have been executed.

The computer therefore can be considered simply as a calculator which has two basic requirements:

(i) a programme of instructions
(ii) numerical data upon which to operate.

Programmes for most computers can be prepared in written form either in machine language (i.e. numerically coded instructions) or in one of the programme languages such as Autocode, Fortran, or Algol (a mixture of words and numbers) according to the computer equipment being used.

The numerical data are prepared in manuscript, usually on a printed form (see Figure 5.13, page 51) in the order required by the programme.

The programme and the data are then typed separately on a special typewriter which, at the same time as it produces a typewritten copy, punches holes in a series of punched cards or in a length of paper-tape in such groups as to represent, in code, the programme of instructions and the data. Magnetic tapes are also extensively used for this purpose. If these cards or tapes are fed into the computer, the computer is capable of detecting the punched holes and reading from them what the programme of instructions and the data are. The computer will then carry out the instructions at immense speed using the data provided, and produce the required answers automatically by punching holes in outgoing cards or tape. These can then be decoded and the answers printed in intelligible 'output' form by passing them into a printing machine (see Figure 5.14, page 54).

The preparation of computer programmes for complex operations can be a time-consuming and expensive task, but once a programme of instructions has been written it can be used in the computer over and over again with different sets of data. Computer firms, universities and others are, therefore, building up 'libraries' of computer programmes for the well-defined computational operations which are likely to be required for repeated use. Amongst these are programmes for analysing networks.

Programmes for network analysis

Programmes for the analysis of networks have been written for many modern computers. They vary in detail and capacity but are all capable of performing the same basic analysis of networks as explained in Section 1 of this chapter.

These programmes are in the main general purpose programmes in the sense that, although they all carry out the same basic analysis, parts of the programme can be passed over, or differences in the arrangement of the output data can be achieved by adding to the input data suitable directives based on the use of the computer's ability to choose between alternatives according to some specified criterion. For example, an instruction or directive might take the form 'if the answer obtained is greater than 100, miss out the next 25 instructions, otherwise carry on with the next instruction'.

The method by which the calculations are carried out is somewhat different from that explained earlier because of the need for the computer to sort out the activities into their sequences, but this is of no consequence to the user. All that is necessary for the user is to know how to present to the computer the data and logic contained in the arrow diagram, and what data he requires from the analysis.

Computer input

The input data for the computer may consist of either a list of the activities (including dummies) in the network if an activity-orientated output is required, or a list of events and activities if an event-orientated output is required. An activity-orientated system is one which concentrates on activities and gives the results of the analysis in the form of a table of activity times similar to that in Figure 5.10,

Computer department **Master files / Test schedules** Job:

```
1 2 3 4 5 6 7 8 9 10 1 2 3 4 5 6 7 8 9 20 1 2 3 4 5 6 7 8 9 30 1 2 3 4 5 6 7 8 9 40 1 2 3 4 5 6 7 8 9 50 1 2 3 4 5 6 7 8 9 60 1 2 3 4 5 6 7 8 9 70 1 2 3 4 5 6 7 8 9 80

AER  1   6   7   6   1300669 5BUILDING TWO BUNGALOWS                          1
 ER  300669                                                                   2
AER  H010969   1                                                              8
AER  H251269   2                                                              8
AER       0    10            SIGN CONDITIONAL CONTRACT FO      10             31
AER       0    10            R PURCHASE OF LAND                               3
AER      80    60            DUMMY                                            3
AER     150   160            BUILD SHELL (A)                   30             3
AER     180   200A           ELECTRICAL WIRING (A)              4             3
AER     280   290            INTERIOR TOPCOAT (A)               3        L    3
AER     185                  CONNECT SERVICES TO MAIN SUP                    K61
AER     185                  PLY                                             K6
AER       0                  START EVENT                                     86
AER     300                  END EVENT                                       E6



 ET  300669                                                                   2
CET      60   115                                              33             3
CET     560   570                                              10             3
```

Reproduced by kind permission of
International Computers Limited, Reading, Berkshire

Figure 5.13

and an event-orientated system concentrates on events and gives the results of the analysis in the form of a table of event times similar to that in Figure 5.4.

Although early computer programmes imposed restrictions on the order in which the activities had to be listed and the way they were numbered, modern computer programmes make no such restrictions. All computer systems do however require the data from the network to be presented on their particular standard printed form, an example of part of one of which is given in Figure 5.13, page 51.

The basic input data required for each activity in an activity-orientated system are:

(i) a brief verbal description of the activity

(ii) the unique pair of events numbers

(iii) the duration.

For an event-orientated system it is necessary in addition to provide for each event:

(iv) a brief verbal description of the event

(v) the event number.

The data can be punched on cards or tape though cards are more commonly used. One punched card is prepared for each event or activity and the bundle of cards so prepared is sufficient to give to the computer a complete description of the configuration of the network and the interdependencies between the activities—clearly the computer will know that for instance all activities ending in event *x* must immediately precede all activities starting from event *x* i.e. 1–4 and 3–4 must precede activities 4–6 and 4–7. Given also all the activity durations the computer can then perform all the basic calculations explained in Section 1 of this chapter.

The number of letters which can be used in describing each event or activity may be limited by the capacity of the storage space in the computer. Most programmes allow between 25 and 40 letters per event or activity description so that this limitation is likely to be an inhibiting factor only in exceptional circumstances.

Besides the basic data for the computer input, various titles and codes relating to the project as a whole are normally required. A number of directives are also required relating to the type of output desired; the number of copies; the order in which the events or activities are to be listed; the project starting date if times are to be given in the output as calendar dates instead of times from project commencement; codes to identify those activities whose data have changed since the previous analysis; departmental codes, so that the activities can be grouped under departments in the output; and so on.

With most computer programmes it is possible to specify target dates for events in the network and any sequence of activities which must be shortened to meet these targets are shown with negative float in the computer output (see page 43).

The use of simple forms reduces the preparation of input data for the computer to a straightforward clerical operation involving only reading directly from the prepared arrow diagram. This can nevertheless be a laborious, time-consuming operation but mistakes can be reduced by taking the events in numerical order and listing for each all the activities originating from it. There is no alternative to carrying this out manually—the computer will neither prepare the arrow diagram nor list the data for analysis, it will only carry out the analysis itself and print out the results.

When preparing data for second and subsequent updating analysis it is not necessary to prepare the whole list of events and activities again. It is necessary only to provide new data relating to those events and activities which have changed since the previous analysis. This may involve no more than changing to zero the durations of those activities which have been completed since the previous run, or estimating the outstanding time required for completion of current activities. A new punched card is prepared for each of these activities to replace the original one in the bundle. The whole bundle can then be reprocessed to yield up-to-the-minute data for management control.

Computer output

The output data obtained from a computer analysis of a network can be arranged in a number of ways. The basic part of the output is however a table of event times and slack, as shown in Figure 5.4, or a table of activity times and floats as shown in Figure 5.10. In the output the events and activities are identified both by the relevant pairs of event numbers and, for convenience, by the verbal description given in the input data, although the verbal description does not of course play any part in the calculations. An example of a part of a computer output is given in Figure 5.14:

I C L 1900 SERIES PERT 10/11/70 OUTPUT SHEET NUMBER 66

PROJECT ER BUILDING TWO BUNGALOWS RUN 2 TIME NOW 30JUN69 PAGE 2

ACTIVITIES BY SCHEDULED START (6)

S/P	PREC EVENT	SUCC EVENT	REPORT CODE	DESCRIPTION	DUR	EARLIEST START	SCHED START	SCHED FINISH	LATEST FINISH	REM FLOAT	RESOURCES R1	R2
AA	120	130	STA	BRICK UP TO DAMP COURSE (A)	.3	24OCT69	25OCT69	29OCT69	29OCT69	.0	2AA .3	2CA .3
AA	430	440	STB	LAY HARDCORE AND CONCRETE OVERSITE (B)	.2	28OCT69	28OCT69	30OCT69	30OCT69	.0	4AA .2	
AA	130	140	STA	LAY HARDCORE AND CONCRETE OVERSITE (A)	.2	28OCT69	29OCT69	31OCT69	31OCT69	.0	4AA .2	
AA	450	460	STB	BUILD SHELL (B)	3.1	30OCT69	30OCT69	21NOV69	21NOV69	.0	2AA 3.1	2CA 3.1
AA	150	160	STA	BUILD SHELL (A)	2.4	30OCT69	31OCT69	19NOV69	19NOV69	.0	2AA 2.4	2CA 2.4
AA	160	170	STA	ERECT ROOF (A)	.2	18NOV69	19NOV69	21NOV69	21NOV69	.0	1AA .2	1DA .2
AA	170	180	STA	FELT AND TILE ROOF (A)	.2	20NOV69	21NOV69	24NOV69	24NOV69	.0	1AA .2	2EA .2
AA	460	470	STB	ERECT ROOF (B)	.3	21NOV69	21NOV69	25NOV69	25NOV69	.0	1AA .3	1DA .3
AA	180	200 A	STA	ELECTRICAL WIRING (A)	.4	22NOV69	24NOV69	28NOV69	28NOV69	.0	1FA .4	
AA	180	200 B	STA	INITIAL PLUMBING	.2	22NOV69	24NOV69	26NOV69	28NOV69	.2	2HA .2	
				TO MAIN SUP. (A AND B)			24NOV69	25NOV..			2GA	
AA	200	210	STA	PLASTER WALLS (A)				..DEC69	5DEC69	.0	1AA .6	2JA
AA	500	510	STB	PLASTER WALLS (B)	.5	2DEC69	2DEC69	8DEC69	8DEC69	.0	1AA .5	2JA .5
AA	210	220	STA	CONCRETE SCREAD (A)	1.1	2DEC69	3DEC69	11DEC69	11DEC69	.0	1AA .1 *AA 1.0	1JA .1 *JA 1.0
AA	510	520	STB	CONCRETE SCREAD (B)	1.1	8DEC69	8DEC69	16DEC69	16DEC69	.0	1AA .1 *AA 1.0	1JA .1 *JA 1.0
AA	145	235	ST	DRAIN LAYING AND SEWER CONNECTION (A AND B)	.3	30OCT69	9DEC69	12DEC69	20DEC69	1.1	3AA .3	
AA	230	240	STA	FINAL PLUMBING (FITMENTS) (A)	1.2	10DEC69	11DEC69	20DEC69	20DEC69	.0	2HA 1.2	
AA	220	240 A	STA	CARPENTRY (A)	1.0	10DEC69	11DEC69	18DEC69	20DEC69	.2	2DA 1.0	
AA	220	240 B	STA	ELECTRICAL FITMENTS (A)	.1	10DEC69	11DEC69	12DEC69	20DEC69	1.1	1FA .1	

Figure 5.14

The events or activities can usually be listed in any particular order required. For example, events can be listed in ascending order of slack so that the critical events appear at the top of the list. Alternatively, they can be listed in order of earliest time or latest time. Activities can be listed in ascending order of total float, the critical activities therefore appearing at the top of the list, or in ascending order of EST, LFT, etc., or in departmental groups as required. Critical activities and activities which have changed since the previous analysis, can have code letters printed against them signifying these facts.

It is possible to have the times in the ouput table printed as calendar dates instead of times from the commencement of the project. If a starting date for the project is provided with the input data and the length of the normal working week is specified, the table of event or activity times will be worked out and printed in terms of calendar dates having allowed for statutory holidays in the computation.

The computer output from a re-analysis does not necessarily have to take exactly the same form as the output from the original analysis, nor need it be complete. The form of output can be adjusted for each analysis to meet the needs of the moment.

Services available at computer bureaux

All the major computer bureaux and firms of consultants offer computer facilities for the analysis of networks. The names and addresses of some of these are given in Appendix 3 at the end of this book. Once the arrow diagram for a project has been prepared, the task of having it analysed by computer requires only the presentation of the input data. It is not expensive, and the output data required can usually be available within 24 hours.

The form in which the input data are required and the scope and variety of output print-outs vary considerably from bureau to bureau. The chosen bureau should therefore be consulted in advance on the user's requirements and how they are to be met. (For further details of the capabilities of computers in resource allocation and cost analysis, as opposed to simple time analysis as described here, see Chapter 7, page 67, Monitoring by computer.)

The ease with which these services can be used and their relative cheapness does not however mean that the analysis of a network should always be carried out by computer. On the contrary there are many projects and circumstances in which manual analysis is to be preferred both on economic grounds and on grounds of convenience.

Section 3 **Manual *vs.* computer analysis**

No hard and fast rules can be made for choosing the method of analysis to be employed. The manual method of direct analysis of the network diagram needs care if errors are to be avoided. When one is analysing by computer the possibility of error is virtually confined to the preparation of the data.

The size of the network is not the sole factor determining the choice between manual and computer analysis, although clearly a very small network of less than about 50 activities would almost certainly rank for manual analysis and any large network of over about 500 activities would almost certainly have to be analysed by computer.

Between 50 and 500 activities the choice may be determined by a number of factors. Ease of access to a computer, and the time it will take before the results are returned, are probably as important factors as any. If the planning office is in a remote area it is less likely that a computer analysis would be of value than if the planning office were in London close to a number of computer bureaux.

The length of the project and the number of and type of analyses likely to be required may also play a part in the choice. If the project plan is not too finely balanced and therefore likely to be stable, only one analysis at the outset may be required, which would hardly justify the clerical effort involved in listing the activities and having punched cards prepared. If on the other hand there is an appreciable degree of uncertainty in the project a number of analyses may be required during its progress and there is therefore a greater case for preparing punched cards and using a computer from the start.

In most cases however cost must be the deciding factor. The analysis of a 400–500-activity diagram may occupy a clerk for a week and even then the accuracy is not guaranteed.

The charges for using a computer are calculated in different ways by different computer firms, but generally comprise a small charge for handling in addition to charges per activity for data preparation (punched cards), computer time for the analysis of the network, and provision of the printed output. The charge for data preparation may be as much as twice the charge for the computer time. Each different form of output is charged for separately.

A typical charge for a 500-activity network, providing for a single time analysis and a single output, might be :

	£
handling	6
data preparation	13
analysis	6
output	3
	28

i.e. somewhat over 5p per activity. Each further output, perhaps listing the activities in a different order, would cost an additional £3 or so. Resource analysis would add a further £13 to the cost, plus perhaps £3 for an additional output.

Subsequent computer runs would of course cost somewhat less than the first one as the charge for the preparation of the data relating to amendments would be less than for the data preparation for the initial analysis run. For a 500-activity network for a project lasting 12 months and requiring half a dozen refining runs to meet management's objectives (see page 45), and half a dozen updating runs as it progresses, the total cost for time analysis only might be as much as £250. Resource analysis might cost up to an additional £100. Clearly in a project involving an investment of several thousands of pounds, with the promise of several thousands of pounds of additional revenue, this is a small price to pay for speedy and accurate analyses, and the provision of regular up-to-date management control information on which to base decisions.

Conclusion

Although the calculations involved in the analysis of networks and the methods of doing them have been explained at length it is important to emphasize that, in practice, this forms a relatively small and unskilled part of the technique as a whole. The preparation of the arrow diagram is the key to success in using network techniques and effort put into this will be handsomely rewarded.

Six Scheduling and resource allocation

The data resulting from the analysis of a network diagram is not an end in itself but is produced for the purpose of assisting in the preparation of a work schedule which will be reliable and result in the best use of the available resources as the project proceeds.

The first step in setting up a network planning system, namely the representation of the interdependencies of the constituent activities in the project by an arrow diagram, ignores any reference to a time scale or to the availablility of resources. Both of these must now be taken into account in preparing the schedule and allocating resources to each of the activities in it.

This is a very complex problem indeed. Achieving a reasonable balance between the conflicting demands of manpower, machines and money productivity is extremely difficult but the use of Network Planning Techniques provides a means, particularly in large and complex projects, of achieving a better balance than is normally attainable by the use of traditional planning techniques.

It is not possible in a work of this size to discuss at length and in detail all the problems and ramifications of scheduling and resource allocation. This chapter is, therefore, confined to an explanation of the use of network analysis data in scheduling and to commenting on some of the problems encountered in resource allocation and resource levelling.

Scheduling

(A) RESOURCES UNLIMITED

The ultimate aim of any project planning system must be to provide those who are managing the project with a timetable or schedule of work which can be readily understood and followed. Such a schedule is normally presented in the form of a bar chart.

When the preparation of a bar chart has been preceded by a network analysis for the project all the data relating to activity times and floats are available to assist in the preparation of the bar chart.

Information relating to manpower, machines and money requirements for each activity may also be available and, if not, can be produced if required.

The natural tendency in preparing the bar chart will be to show all the activities starting at their earliest starting times as determined by the network analysis, but there is no fundamental reason why the activities should all be carried out as early as possible. Indeed in order to defer spending capital till the last possible moment and thereby minimize interest charges on capital, all activities might be postponed to their latest starting times, although this would of course mean that an unforeseen delay in any one activity would delay the completion of the project.

Knowledge of the limits within which each activity must be scheduled can be of great value in preparing the bar chart because alternative timings beyond these limits need not be considered, or at least need be considered only if a delay to the project completion is acceptable or inevitable.

The process of scheduling is, therefore, to find within the limits of total and free float available some efficient combination of individual activity schedule times, which will meet the project objectives in regard to target completion date, any other targets imposed by other departments, and so on. These limits in effect introduce a time restriction into the planning system for the first time but efficiency in this context, though it may imply a measure of resource levelling, does not yet imply any restriction on the availability of resources.

(B) RESOURCES LIMITED

If for the schedule prepared within these time limits the necessary resources cannot be provided at the required times, this amounts to a limitation on the availability of resources and will mean that it may not be possible to achieve in practice the end date calculated in the analysis of the arrow diagram. The schedule must be revised to bring the level of demand for each resource within the limit of availability of that resource at all times. This may result in an estimate of the project completion date later than the original calculated from the analysis of the arrow diagram.

It should be made clear, however, that there is no discredit in a revision of the completion date in this way—it is a direct consequence of making it part of the scheme of things to base the calculations in the network stage on the assumption that resources will be available in unlimited amounts, and subsequently making the necessary adjustment to allow for the fact that in practice this may not be so. In this way the essential choice facing the manager is exposed : whether to employ more resources or accept a later completion date; and the manager now has the facts upon which to make a sound decision. In most cases where the purpose of the project is to provide substantial new revenues, he will find that the benefits from advancing the completion of the project far outweigh the cost of additional resources.

Resource levelling

Whether resources are limited or unlimited, any project schedule representing some combination of individual activity schedule times, will produce for each resource a demand varying with time. By rescheduling individual activities it may be possible to reduce the peaks in demand or otherwise alter the fluctuations to meet some objective. It will seldom be possible to achieve a completely steady demand for a particular resource throughout the project, nor will it often be possible to find a schedule for which the individual utilizations of all resources are all maximized at the same time. It is more usual to find that the benefit from maximizing the utilization of one resource has to be balanced against the cost of reducing the utilization of another resource and perhaps in some cases the cost of delaying the completion of the project.

There is no alternative in practice to an *ad hoc* approach to this problem. Small projects involving few resources can be handled on a completely 'trial and error' basis. For larger projects however situations will frequently arise in which more than one activity could be given the same resources at the same time. It is necessary therefore to have some rules for deciding which shall have priority; for example whether the activity which is already in progress shall have priority, or alternatively the activity with the lowest float, or the longest duration, or whatever. Clearly a wide variety of combinations of such priority rules can be devised and used for different projects in different circumstances. Unfortunately no single combination of such rules will necessarily give the best solution to this problem in all cases—the schedule produced is considerably influenced by the priority rules used. What is possible, however, is to set up a number of different schedules using different priority rules and, by examining the resource demand graphs, target dates, etc., resulting from each of these, suggest further adjustments to the schedule, calculate new resource demand graphs for examination, and so on. By such an iterative process it is possible to produce a better schedule for a large and complex project than is possible by traditional methods.

What is in effect being done is that the planning (placing activities in the right order in the network) is being kept separate from scheduling (in the sense of programming to a time scale). The consequences of limitations in the availability of resources, including time constraints resulting from the imposition of target dates etc., are not considered until the scheduling stage which follows, and is quite distinct from, the planning stage. At the scheduling stage, in order to smooth out fluctuations in the demand for resources, decisions are made to programme some of the activities at times later than their EST's. These decisions can be built into the network using 'time only' arrows (see page 13) to represent delays decided upon. The re-analysis of the amended network will therefore give a project completion time and a table of activity times which take these decisions into account. This, along with various other statements on resources, represents the plan for the project.

There is an enormous amount of computation and data processing involved in a large-scale resource levelling exercise. This work, and the production of the statements and reports which represent the plan for the project, can be transferred to a computer. The range and scope of possible computer outputs are outlined in Chapter 7.

Example

In order to clarify and emphasize some of the mechanics of scheduling and resource levelling a very simple example is discussed in the ensuing pages.

(A) SCHEDULING

The bar chart in Figure 6.1 for the demonstration network used in the previous chapter (Figure 5.2) has been prepared initially with all the activities starting at their EST's. The durations and floats have been taken from the table in Figure 5.10 and represented by different types of shaded bar. The activities have been listed in ascending order of total float so that the critical activities appear at the top. Only one resource (men) is considered.

In the column to the left of the bar chart the number of men required for each activity is shown. Below the bar chart is shown the

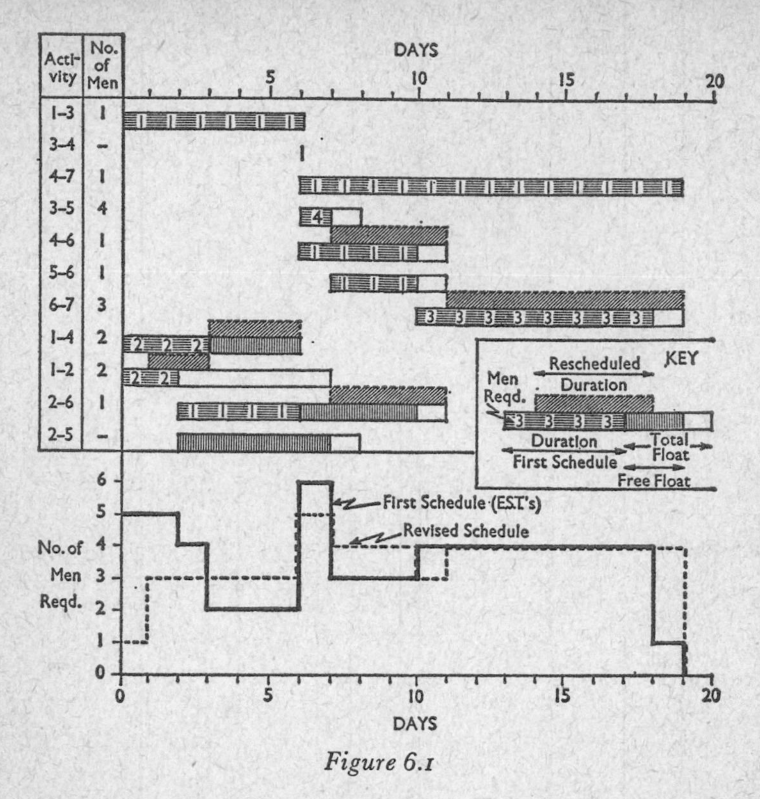

Figure 6.1

graph of the variations in demand for labour throughout the project resulting from totalling the labour requirements day by day.

This bar chart as it stands represents a feasible schedule for carrying out this project in 19 working days. It assumes, however, that in effect resources are available in unlimited amounts and that the project manager can be supplied with widely differing numbers of men each day, or is prepared to accept a low labour utilization by maintaining a constant strength of say five men on the project throughout, with the need to provide a sixth man on the seventh day. Neither is particularly desirable from the point of view of efficiency.

(B) RESOURCE LEVELLING

It is now possible to examine the scope for resource smoothing by rescheduling some or all the activities within the limits of float available. Up to this stage it has been implied that resources are always available in unlimited quantities. Now it is necessary to take into account any limitations which may exist in the availability of resources, including any restriction in time implied by setting a target date for completion of the project.

The only logical way to tackle this problem is to examine the possibilities of moving activities away from those periods when demand for resources is high to those periods when demand for resources is low.

In the example in Figure 6.1 there are two peaks in demand for resources, one lasting the first three days, and a second on the seventh day. Slack demand occurs from the fourth to the sixth day, from the eighth to the tenth days, and on the last day.

By taking advantage of the free float associated with activity 1–4, by using one day of the total float associated with activity 1–2 to finish activity 1–2 immediately before activity 1–4 starts, and by scheduling activity 2–6 at its latest starting time, it is possible to reduce the demand for labour to one man on the first day and three men from the second to sixth days inclusive. This gradual build up at the beginning of the project is a much more desirable state of affairs.

If advantage is taken of the one day total float associated with activity 4–6 which, along with the rescheduling of activity 2–6 means that activity 6–7 must be delayed one day, then there will be a demand for four men from the eighth day onwards. On the seventh day, however, five men will be required.

These rescheduled activities are indicated by diagonal hatched bars in Figure 6.1 and the resultant resource demand graph, represented by a dotted line, indicates that the project can be carried out by a gang of three or four men, with the addition of a fifth on one day.

It should be noted in passing that by definition advantage can be taken of free float associated with an activity without affecting the scheduling of any other activity. Total float however, as mentioned on page 41, is associated with a sequence of activities between two points on a controlling path. When therefore advantage is taken of the total float associated with an activity to delay that activity, subsequent activities in the same chain in the network will automatically be delayed from their EST's by the same amount and thereby lose that amount of total float. In other words total float can be used only once in a chain of activities and, once it has all been used, any further delay in the chain will delay the completion of the project.

It should also be observed that float associated with a dummy has no meaning in practice. A dummy will carry the same total float as the other activities in the same chain, but this signifies only a range of times at which the dummy (itself a point of zero duration) can occur.

Free float associated with a dummy should be examined to see if it can be associated with another activity. When an event has only dummies with free float originating from it, the smallest of the free floats should be added to the activities terminating in that event. Free float associated with a dummy in other circumstances has no practical value or significance.

Where an activity float is very large the possibility of reducing the resources allocated to it and thereby increasing its duration, should not be overlooked. It will not always be possible or economic to do this. For instance, it would not normally be economical to replace a mechanical digger by a squad of men to carry out an excavation just because sufficient float is available to do so.

The schedule prepared for the example in Figure 6.1 is not the only one possible. For example, activity 1–4 could be returned to its earliest starting time and activity 1–2 started after 1–4 was completed. These changes would have further effects on the resource demand graph, but none, in this particular example, will remove the need to have five men present on one day.

This raises an interesting point. The Project Manager now has the choice of arranging for a fifth man to be present on one day, or delaying the project by one day if only four men can be made available. A delay of one day to the critical activity 4–7 so that all four men could be assigned to activity 3–5 on the seventh day would produce an equivalnet delay to the project. The Project Manager must weigh the cost of providing a fifth man for one day against the cost, including any loss of potential revenue, which would be incurred by delaying the project one day.

The lesson to be learned here is that the project completion date forecast from the analysis of the network may be untenable when the necessary resources are not available. If two or more activities are competing for the same resources at the same time there may be no alternative to accepting a later project completion date than originally forecast by the arrow diagram alone. On the other hand, as already mentioned on page 60, if the necessary resources can be made avalaible the cost of doing so will very often be much less than the benefits to be gained from advancing the completion date of the project.

Exercise

9 Assuming that in Exercise 7 the manpower requirements for each activity are as follows, construct a bar chart showing how the total demand for men varies day by day if all activities start as early as possible.

Activity	N	A	C	B	D	L	J	F	H	K	G	E	P
Men required	5	2	1	3	1	1	2	1	1	1	2	4	5

If a maximum of 5 men can be made available on any one day, will it still be possible to complete the project in 22 days or will it be necessary to accept a later completion?

Conclusion

Although the simple example in this chapter can be used to illustrate some of the problems and difficulties of scheduling and resource levelling, projects in practice are never likely to be so simple. Not only will the network be larger but the planner will be endeavouring to strike a satisfactory balance between the conflicting demands of a number of resources and objectives. The requirements and availability throughout the duration of the project, of several different classes of labour, of several different types of machines, of time and of money, would all have to be weighed in the balance if an elegant solution of the scheduling problem is to be sought. And, because a schedule which may be ideal from the point of view of the utilization of one resource or the achievement of one objective may reduce the utilization of another resource or jeopardize the achievement of some other objective, automatic optimization is extremely difficult, if not impossible. The absence of a set of decision rules to guide a computer, or the planner for that matter, in deciding priorities, makes a trial and error approach to scheduling inevitable. Preparation and analysis of a network does however provide important data for this purpose, and laborious routine calculations can always be transferred to a computer. For an indication of the capability of computers in resource allocation see Chapter 7, page 71.

The plan or programme emerging from this process in the form of various time and resource analyses will not necessarily be the best nor the only feasible plan for implementing the project. There is no single plan which is right and all the others are wrong. Different experts will put different interpretations on standard practice and procedure, producing different networks for the same project. Different managers will make different amendments in the trial and error process of scheduling, producing different programmes. The planning process has not, nor will it, absolve management from exercising its function. Nor should the possibility that further improvements to the plan could be found inhibit management from proceeding to implementation on the basis of the plan now prepared. 'The best is the enemy of the good.'

The use of the plan in monitoring progress and management control is the subject of Chapter 7.

It should be emphasized in conclusion that great value has been derived from the use of Network Planning Techniques using only the simplest and crudest forms of resource levelling. What is so often of the utmost importance is that the project should be completed at the earliest possible date because loss of revenue on capital equipment is involved while the project lasts. Whether the project be the construction of new plant where the money spent in the early part of the project cannot bring in revenue until the project is complete, or the overhaul of existing capital equipment during which normal revenue-earning ceases, the cost of loss of revenue in many cases far outweighs the cost of providing the resources required to advance the completion date. In setting up the initial plan for the project therefore a substantial pay-off can accrue from directing maximum effort to providing the resources necessary to minimize project time.

Seven **Monitoring and control**

The foregoing chapters have dealt with the preparation of a plan or programme for the project. There are clearly immense advantages in doing this as early as possible and certainly before the implementation of the main part of the project begins.

However much time and effort are expended on setting up the plan not all predictions will be correct nor will all contingencies be foreseen. As the project proceeds it will transpire that some estimates of duration were in error, unpredictable delays will affect other durations, some logical relationships will not have been quite accurately represented, suppliers will default on delivery dates, absenteeism or strikes will reduce the availability of resources, and so on. To remain dynamic the network planning system must provide a means of quickly evaluating the effects of such contingencies, monitoring the progress of the project and simplifying management control.

Evaluating the effects of contingencies

If an unforeseen contingency arises which causes the completion of an activity to be delayed beyond its planned date then, so long as the delay to that activity is less than the float remaining available to it in the plan, there will be no direct effect on the completion date of the project although some replanning might be required if the delay upsets a finely balanced resource schedule.

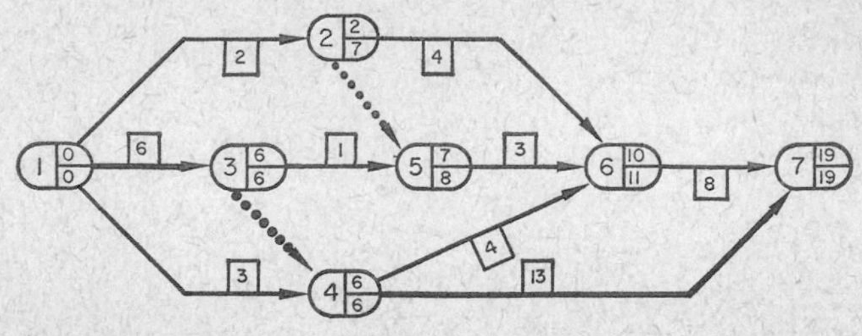

Figure 5.2

Returning to the example used for demonstration purposes in Chapter 5 (Figure 5.2 repeated here for convenience) the critical path of 19 days comprised activities 1–3, 3–4, 4–7. If, through some unforeseen contingency, activity 3–5 is delayed 1 to 2 days, a second critical path through 1–3, 3–5, 5–5, 6–7, also of length 19 days, is created, but the length of the project is not affected. If however activity 3–5 is delayed by a further 1 to 3 days then path 1–3, 3–5, 5–6, 6–7, is 20 days long and becomes critical on its own. The original critical path 1–3, 3–4, 4–7 is then no longer critical activity 4–7 now having a float of 1 day because the project completion is delayed by one day.

In these circumstances however, because the Project Manager is quickly made aware that a new critical path has appeared, it is possible to consider allocating more resources to activities 5–6 or 6–7 to reduce one of them by one day and thereby restore the project to 19 days. Adjusting the plan following an unforeseen contingency is in effect re-planning and the methods open to the manager for doing this are the same as for adjusting the network in the planning stage to enable specified objectives to be met (see page 45).

Although unforeseen contingencies do result in different sequences in the network becoming critical, there are many projects in practice in which, once the real critical path has been identified, the same activities remain critical throughout the project duration. In many projects most of the activities have large floats so that it requires a substantial delay to a non-critical activity to cause a shift in the critical path.

Delays when they do occur do not necessarily require a complete updating re-analysis of the network. Only when an activity is delayed by an amount greater than its total float, and it therefore becomes critical, will it be necessary to examine subsequent activities on the new critical path for the possibility of reducing their durations and thereby restoring the original planned project completion date. A re-analysis of the network, along with re-programming the resources, might then be undertaken if necessary to evaluate the effects of accumulated changes and, on the results of the re-analysis, a decision made on any remedial action required.

Communication

Before the progress of the project can be monitored a feedback of progress information is required from the field of operation. It is not within the scope of this book to dwell in detail on the clerical procedures required to do this. Suffice it to say that what is required is something simple, at intervals, comparing, activity by activity, the information fed to the field managers from the network plan, with the actual situation at the reporting date.

This raises a much-debated issue : how much information, particularly in regard to activity floats, should be issued to field managers and supervisors from the network plan? There is of course everything to be gained, once personnel have been trained to a full understanding of the technique, from communicating to them, or at least making available to them, all the information from the network which could have a relevance to their work and decisions. It is as well to make it clear however that float associated with an activity cannot always be made readily available, particularly as this might affect other activities or upset the resource programme. The field manager has a responsibility to meet the *programmed* date for all his activities whatever float they may have.

The field managers are therefore periodically reporting progress of each activity against a programmed date, and the consumption of resources against estimated targets. The Planning Office is then able to put all these reports together and monitor the current position of the project.

The Planning Office will not however automatically re-analyse the network at the end of each reporting period. The re-analysis of the network would produce new programme dates for all the activities and it is better, unless it is absolutely necessary, to avoid issuing new programme dates to field managers too frequently lest the Planning Office be charged with not being able to make up its mind.

Monitoring—manually

Whether or not the project is re-analysed at each reporting period the feedback of progress information from the field of operations enables the Planning Office to monitor the progress of the project and prepare the data necessary for management decision and control.

Monitoring involves some means of assembling the feedback information and evaluating its significance to the project. For some projects this can be done by manual methods based either on the network itself or on the table of activity times.

The main information required is

(i) a list of those activities which have been completed
(ii) for those activities currently in progress an estimate of the time required for their completion
(iii) an y variation in starting dates and durations of outstanding activities from planned dates and times.

The information can be recorded on the network itself in the form of revised activity durations (completed activities are given zero durations), and a cursor line added cutting each arrow representing a current activity at a point a proportionate distance along the arrow. An example is given in Figure 7.1 based on a review at the eighth day of the plan in Figure 6.1, page 62, for the network in Figure 5.2, page 67:

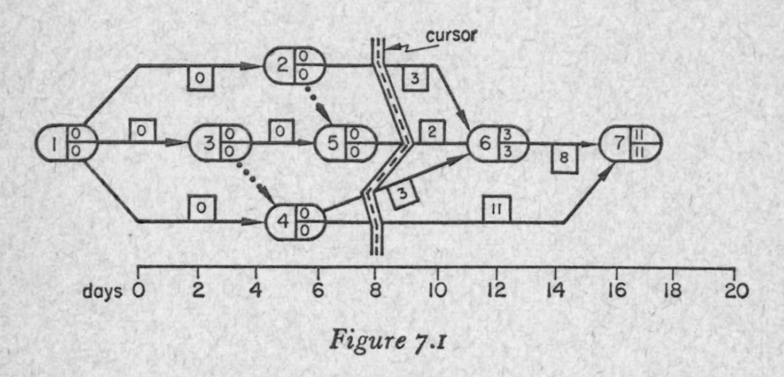

Figure 7.1

With the revised durations an analysis in the normal way will give the completion date (in this case 11 days after the eighth day) and activity floats (for example activity 5–6 has TF and FF each of 1 day).

The current state of progress is thus fully monitored.

Monitoring—by computer

Clearly the scope for manual methods of monitoring is limited to relatively small projects. Networks of a few hundred activities can be successfully analysed by manual methods, but when resource analysis is required in addition the maximum size of network which can be successfully monitored without a computer is very small.

Computer programmes of immense power and versatility are now available to carry out the huge volumes of data processing and calculations associated with thorough time and resource analysis of large and complex projects. Anything but the briefest review of the capabilities of these programmes is however beyond the scope of this book.

Briefly what these programmes are designed to do is to provide the data processing and reporting facility for all network analysis operations. These include:

(i) time analysis as already described in Chapter 5, Section 2

(ii) resource scheduling according to specified sets of priority rules as described in Chapter 6; modern computers are capable of handling very complex situations involving:

(a) up to perhaps 100 different resources with each activity employing different levels of several of these throughout its duration
(b) the possibility of specifying different levels of availability from time to time of each resource
(c) the possibility of stopping work on an activity and restarting when convenient
(d) the possibility of modifying the priority rules to meet special requirements, and so on

(iii) cost analysis linked with the time analysis; it is customary for this to be done on the basis of groups of activities rather than individual activities and linked to an existing cost code structure

(iv) the ability to analyse a part (sub-net) of the total network as an entity in itself

(v) the ability to deal with activities of uncertain duration as for example research and development activities; this is the basis of the PERT system referred to in Chapter 1 and briefly described in the appendix at the end of this book

(vi) the ability to provide a wire variety of reports according to the needs of the user, including:

(a) tables of activity times and floats, arranged in different groups, and listed in any specified order with activity times shown in terms of elapsed time from the start of the project, or in calendar dates as already outlined on pages 53–5;
(b) bar charts showing each activity in its planned position against a time scale

(c) summary reports for higher levels of management based on a relatively few key activities and events
(d) resources utilization reports
(e) for a suitable period into the future sectionalized reports suitable for issue to field managers and supervisors.

(vii) the ability to deal with a multi-project situation.

Examples of some computer input forms and printed outputs are given in Appendix 2.

The computer programmes available for these purposes vary considerably in detail. Before attempting to use them therefore the potential user is recommended to discuss his requirements in relation to programme capabilities with a computer manufacturer or consultant. Some computer bureaux are listed in Appendix 3.

Management control

No management technique, however elegant and sophisticated, will absolve Management from the need to exercise control through making decisions. Management techniques will however, by providing the relevant information, enable management to make better-informed decisions and thereby exercise a finer degree of control than would otherwise be possible.

The network plan which has evolved acts as a base against which to monitor progress. When things go wrong the Project Manager can, using a modern computer, promptly assess the significance of the changing circumstances and quickly decide how to restore the situation; whether to use this new method instead of the method originally proposed; whether to use a machine instead of manual labour; whether to prefabricate instead of constructing *in situ*; whether to delay this expenditure or advance it; what acceleration in the programme is required to compensate for the failure of a supplier; how far he can risk proceeding whilst waiting for planning permission; whether a contractor can be allowed an extension of time or be obliged to adhere to the planned date.

These are the kinds of decisions the Project Manager will be called upon to make to maintain control of the project. The monitoring of progress through the use of critical path planning techniques will enable him to become aware at the earliest possible moment of the problems and contingencies which inevitably arise, and to deal with them promptly and efficiently. In this way he will be able to direct the course of the project more assuredly towards the planned completion date and the achievement of Management's objectives.

Eight An application to a civil engineering project

In order to demonstrate the use of Network Planning Techniques in practice this chapter is devoted to a description of their use in a typical railway civil engineering project. The fact that the project is fictitious is of no significance. This project was prepared as a training exercise and has been chosen for presentation here for the sake of simplicity and because it is fairly typical of modest projects to which network techniques can be applied. Through it, it is possible to demonstrate the methods of dealing with many of the snags which arise in practice. The exercise also has the merit that as far as possible the civil engineering problems in the project have been reduced to their simplest terms so as to be readily understood by the non-engineer.

The project

The four-track railway line between Blandford Grove and Westroop stations is not straight but follows a fairly wide loop (see Figure 8.1). It is proposed to replace the existing loop by four straight tracks between the two stations.

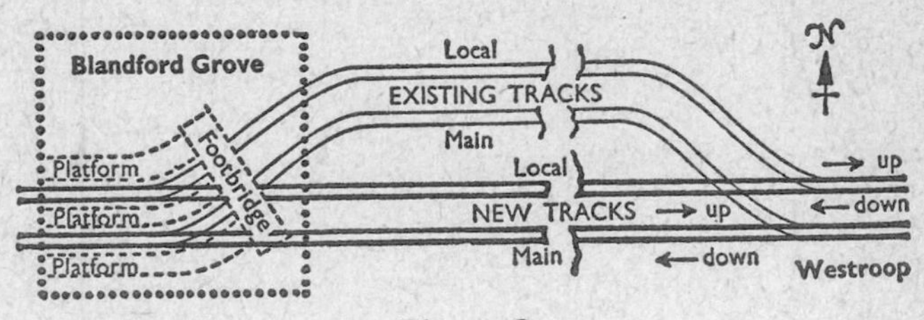

Figure 8.1

Blandford Grove station and the tracks in the vicinity are in a cutting and the new tracks in this vicinity will also be in a cutting.

Although the project will include signal engineering work and some mechanical and electrical engineering work, it is clear that these can easily be planned to fit in with the civil engineering work which forms the major part of the project. It is clear also that the work of laying the new tracks westward from Westroop station will be a relatively straightforward repetitive operation and the complex part of the project will be in the vicinity of Blandford Grove station itself, where a new cutting must be excavated, the station platforms straightened, and a new footbridge built over the track, all with the minimum of interference to existing passenger and rail traffic. It is therefore decided to concentrate on planning the civil engineering operations in the vicinity of Blandford Grove station, the remainder of the work to the eastward having been planned by traditional methods for completion in advance.

The arrow diagram

From an examination of the site and of the plan (Figure 8.2) of the area around Blandford Grove station a method for carrying out the

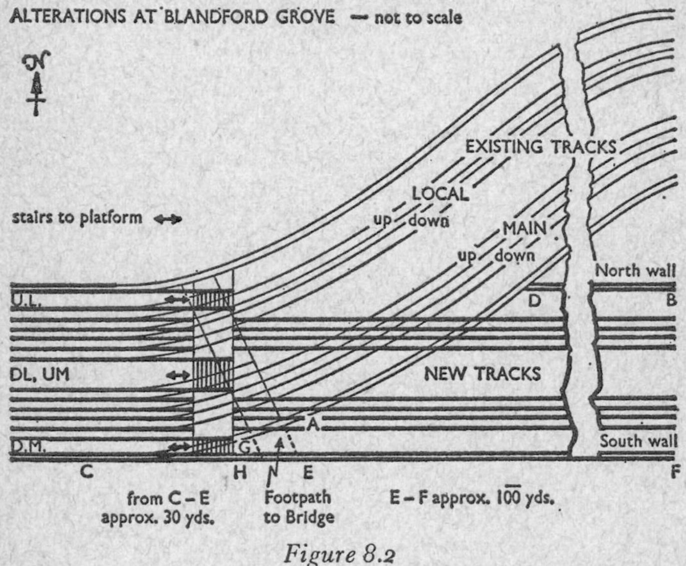

Figure 8.2

project is decided and represented by the arrow diagram (Figure 8.3, page 81). This arrow diagram is prepared by the Civil Engineer in charge of the project.

Because access is provided from the east, excavation of the new cutting can start straight away. It is decided to split the excavation into two parts, area AEFBD and area CAE. The excavation of area AEFBD must precede the construction of both the new retaining walls on the north and south sides of the new cutting and also the demolition of the old retaining wall, AD. The excavation of AEFBD will provide access to area CAE and must therefore precede the excavation of area CAE.

It is decided to scrap the old footbridge which gives access to the station platforms and erect a new one, but in the meantime a temporary bridge must be constructed at the west end of the station. This temporary bridge must be open for the use of passengers before the excavation of area CAE since the path giving access to the old bridge crosses this area. This does not, however, mean that the demolition of the old bridge need precede the excavation of area CAE—the demolition of the old bridge need only precede the demolition of the centre and up-local platform ends and the retaining wall CA, on which the old bridge rests. But after excavating area CAE it is considered necessary that the construction of the new retaining wall CE must precede the demolition of the old wall CA and the down-main platform because of the risk of a landslide on to this platform.

Since both the north and south retaining walls are fairly long, building and backfilling are shown as overlapping activities, the second part of wall building proceeding concurrently with the first part of backfilling (see page 10). Backfilling of all walls need be completed only by the end of the project.

Laying new main tracks and laying new local tracks are shown following the construction of the new south and north walls respectively so that there can be no risk of landslides on to newly laid track.

The installation of the new main tracks up to the junction with the old tracks must be preceded by three activities : the construction of the south wall, the demolition of the old retaining wall AD and shortening the down-main platform. Similarly the installation of the new local tracks up to the junction with the old local tracks must be preceded by the construction of the new north wall, the demolition of the centre platform end, and by connecting up the new main tracks (which includes removing the old main tracks running across the path of the new local tracks).

Foundations for the new bridge trestles must follow the demolition of the corresponding platform ends, and in the case of the centre platform and the up-local platform these foundations must also follow removal of the appropriate lengths of old main and local tracks respectively, when the new tracks are connected.

While these concrete foundations are hardening, the platforms can be constructed round them using vertical concrete facing blocks, backfilling the body of the platform with rubble and covering with paving stones. Following the cessation of traffic on the old local tracks when the new local tracks are connected, a bedstone can be laid in the old north retaining wall without the need to provide screens to protect passing trains. This bedstone will support the end of the new footbridge and when it has hardened, and the three trestles have been erected on the hardened foundations at the platform ends, the bridge spans can be erected.

The temporary bridge must then be removed in order to complete the project.

The analysis of the arrow diagram

The activity durations are estimated in working days assuming that the most economical method is used under 'normal' conditions and that the necessary manpower, machines and money are available. Some of the durations (see Figure 8.3, page 81) are affected by special circumstances. For example the demolition of the centre platform end takes 14 days. This is longer than might appear necessary because the spoil has to be removed by barrow, there being no access for motor lorries. The demolition of the old retaining wall AD with a duration of 30 days may also seem unduly time-consuming but this is accounted for by the need for special care when working close to passing trains.

The analysis of the arrow diagram results in an estimate of 99 working days for the project duration. It also brings to light the fact that this duration is governed by two almost equally critical paths. One path (events 1–4–5–6–7–11–12–13–17) comprises all the excavation and the other work in the vicinity of the down-main platform, the other (events 4–8–9–17) comprises the construction of the south retaining wall. The former path is critical (total float zero) and the latter path has a total float of one working day. These two paths merge into a single critical path (events 17–19–20–21–22–23–24–25–26–27–34–35–36) which comprises the work involved in laying and connecting both local and main tracks, and the work on the up-local platform end.

The earliest and latest starting and finishing times and the total and free floats for all the activities are shown in Table 1, pages 78, 79.)

Adjusting the programme to meet management's objectives

The predicted length of the project depends on the estimates prepared for the critical and near critical activities. These estimates are therefore refined and revised until accuracy can be reasonably guaranteed.

The arrow diagram and the results of its analysis are now used as a means of communicating well in advance to all concerned the best estimate of the project completion date and the activities which govern it.

While the platform ends are being reconstructed their effective length will be reduced and the trains will have to be 'double-stopped'. The Traffic Manager points out that this could have serious effects on passenger business if continued into the summer period. It is decided therefore that platform reconstruction must be completed before the summer timetable comes into operation and that the plan must be adjusted to meet this management objective.

In the period between the agreed project starting date, 24th February 1964 fixed by the completion of work to the eastward, and the introduction of the summer timetable on 15th June 1964 there are 81 working days including four Sunday 'possessions'.* It is agreed that during these Sunday 'possessions' the traffic on the tracks can be diverted elsewhere to allow the engineers to take possession of the track in order to carry out the following operations:

1 erection of temporary bridge (last day of 8-day duration)
2 demolition of old bridge (last day of 4-day duration)
3 connect main tracks (1 day)
4 connect local tracks (1 day).

This means therefore that events 32, 27 and 30 must be reached in 81 days at the latest instead of 95 days at the latest as in the first

* 'Possessions' is railway parlance for those occasions when the Traffic Manager agrees that rail traffic on a particular line shall cease for a specified period to enable the engineers to 'take possession' of the track for maintenance or other engineering works; these arrangements are often made for a Sunday when traffic would in any case be light.

Table 1 Alterations at Blandford Grove Station

ANALYSIS OF NETWORK – ORIGINAL

UL = Up Local DL = Down local UM = Up main
DM = Down main P = Possession (one day)

Activity Description	Event Nos.	Duration	Earliest Start	Earliest Finish	Latest Start	Latest Finish	Total Float	Free Float
Build temporary bridge	1–2	8	0	8	12	20	12	0
Excavate area AEFBD	1–4	20	0	20	0	20	0	0
Demolish old bridge	2–3	4	8	12	36	40	28	0
Dummy	2–5	0	8	8	20	20	12	12
Dummy	3–11	0	12	12	40	40	28	28
Break into DL, UM platform	3–20	14	12	26	60	74	48	48
Break into UL platform	3–24	2	12	14	80	82	68	68
Dummy	4–5	0	20	20	20	20	0	0
Build 30 yds. S. wall EF	4–8	15	20	35	21	36	1	0
Build 30 yds. N. wall DB	4–14	15	20	35	39	54	19	0
Demolish old wall AD	4–17	30	20	50	36	66	16	16
Excavate area CAE	5–6	5	20	25	20	25	0	0
Build retaining wall CE	6–7	15	25	40	25	40	0	0
Dummy	7–11	0	40	40	40	40	0	0
Backfill retaining wall CE	7–36	7	40	47	92	99	52	52
Build 70 yds. S. wall EF	8–9	30	35	65	36	66	1	0
Backfill (30 + 50) yds. S. Wall EF	8–10	20	35	55	74	94	39	10
Dummy	9–10	0	65	65	94	94	29	0
Dummy	9–17	0	65	65	66	66	1	1
Backfill 20 yds. S. Wall EF	10–36	5	65	70	94	99	29	29
Demolish old wall CA	11–12	20	40	60	40	60	0	0
Shorten DM platform	12–13	6	60	66	60	66	0	0
Dummy	13–17	0	66	66	66	66	0	0
Foundations DM bridge trestle	13–18	6	66	72	82	88	16	0
Build 50 yds. N. wall DB	14–15	20	35	55	54	74	19	0
Backfill (30 + 30) yds. N. wall DB	14–16	16	35	51	78	94	43	4
Dummy	15–16	0	55	55	94	94	39	0
Dummy	15–21	0	55	55	74	74	19	19
Backfill 20 yds. N. wall DB	16–36	5	55	60	94	99	39	39
Lay main tracks	17–19	7	66	73	66	73	0	0

continued on next page

continued

Activity description	Event Nos.	Duration	Earliest Start	Earliest Finish	Latest Start	Latest Finish	Total Float	Free Float
Harden foundations DM bridge trestle	18–31	7	72	79	88	95	16	0
Face, backfill and pave DM platform	18–32	7	72	79	88	95	16	0
Connect main tracks	19–20	1	73	74	73	74	0	0
Dummy	20–21	0	74	74	74	74	0	0
Foundations DL, UM bridge trestle	20–28	6	74	80	81	87	7	0
Lay local tracks	21–22	7	74	81	74	81	0	0
Connect local tracks	22–23	1	81	82	81	82	0	0
Dummy	23–24	0	82	82	82	82	0	0
Lay bedstone in N. wall	23–33	2	82	84	88	90	6	0
Foundations UL bridge trestle	24–25	6	82	88	82	88	0	0
Harden foundations UL bridge trestle	25–26	7	88	95	88	95	0	0
Face backfill and pave UL platform	25–27	7	88	95	88	95	0	0
Dummy	26–27	0	95	95	95	95	0	0
Erect trestle and stairs UL platform	27–34	2	95	97	95	97	0	0
Harden foundations DL, UM bridge trestle	28–29	7	80	87	88	95	8	0
Face, backfill and pave DL, UM platform	28–30	8	80	88	87	95	7	0
Dummy	29–30	0	87	87	95	95	8	1
Erect trestle and stairs, DL, UM platform	30–34	2	88	90	95	97	7	7
Dummy	31–32	0	79	79	95	95	16	0
Erect trestle and stairs DM platform	32–34	2	79	81	95	97	16	16
Harden bedstone	33–34	7	84	91	90	97	6	6
Erect bridge over tracks and to N. wall	34–35	1	97	98	97	98	0	0
Remove temporary bridge	35–36	1	98	99	98	99	0	0

plan in Figure 8.3, page 81. In cases of large, complex projects it will usually be desirable to impose such target dates on the events concerned and then re-analyse the network to identify cases of negative float (see page 43), some of which might not be on the critical path. In this relatively simple network however the critical and near critical activities have already been identified and the steps towards meeting the Traffic Manager's target dates would probably be as follows.

Because there is only one day's total float for building the south wall (path 4–8–9–17) any reduction in the duration of the activities in path 4–5–6–7–11–12–13–17 reduces the length of the project by only one day unless the time to build the south wall is also reduced by the same amount less one day. The engineer in charge therefore seeks more economical reductions elsewhere.

The excavation of area AEFBD (activity 1–4) is holding up the start of both the critical and the sub-critical paths and a reduction in this activity would allow both to start earlier. This would reduce the length of the project by a like amount since the only activity in parallel with it, namely building the temporary bridge (activity 1–2), has a total float of 12 days.

The excavation of this area could be accomplished in a shorter time by bringing to the site more excavators and increasing the fleet of lorries for disposal of the excavated soil. The object however is only to shorten the part of this excavation which is on the critical path so that the critical excavation of area CAE (activity 5–6) and the sub-critical construction of the south wall (activity 4–8) can both start earlier. This can be accomplished without increasing the excavating gang by excavating first only the southern half of area AEFBD in, say, 10 working days. This allows both the excavation of area CAE and the construction of the south wall to be started 10 days earlier. The 10-day excavation of the north half of area AEFBD will then follow the excavation of the south half but must still precede the demolition of the old wall AD and the construction of the north wall. The reconstruction of the arrow diagram to take account of these changes in the plan introduces another event, 4A, as shown in Figure 8.4, page 82.

The project length is now reduced by 10 days.

In a similar way the whole seven days of laying of the local tracks (activity 21–22) is shown in the original arrow diagram (Figure 8.3) as being preceded by connecting main tracks (activity 19–20), by breaking into the centre platform end (activity 3–20) and by the

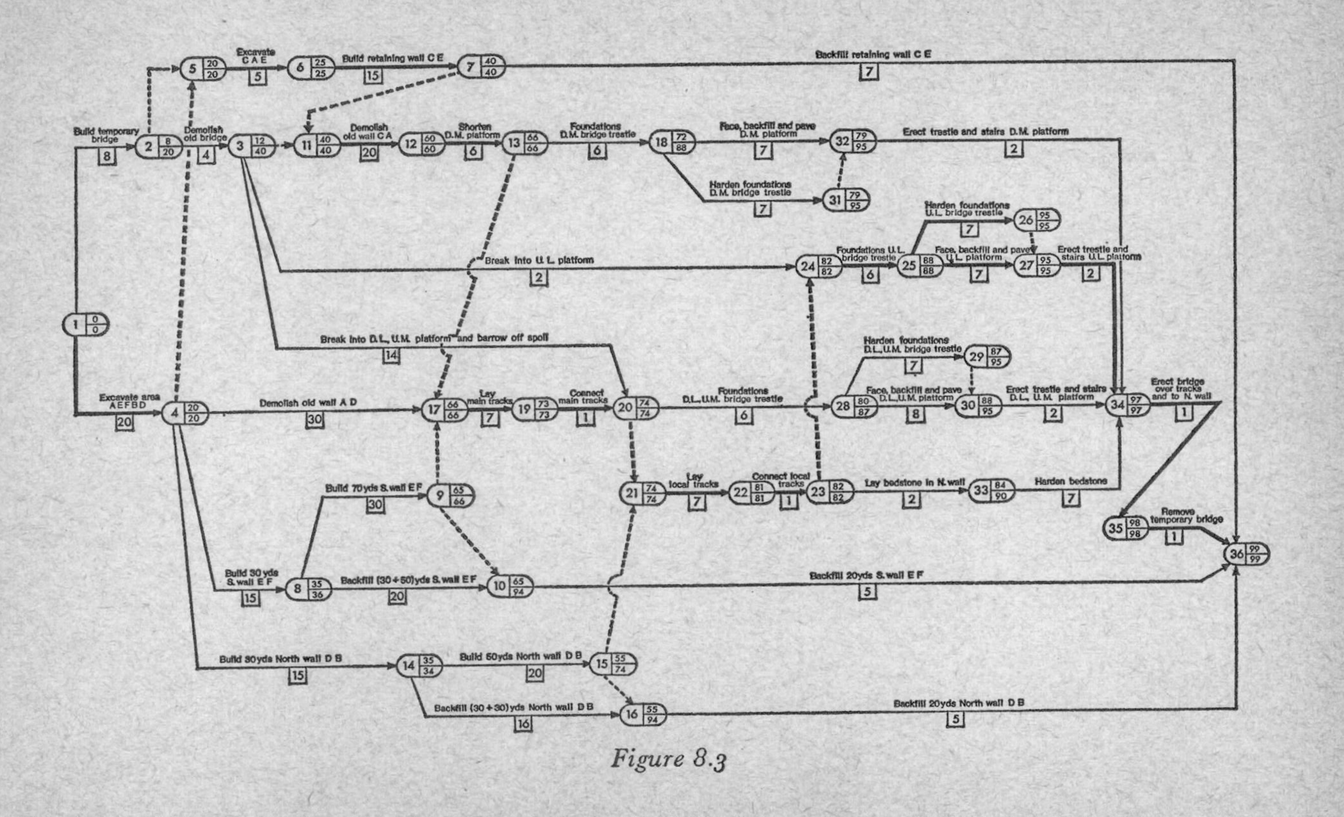

Build temporary bridge
Excavate area AEFBD
Demolish old bridge
Excavate CAE
Build retaining wall CE
Backfill retaining wall CE
Demolish old wall CA
Shorten D.M. platform
Foundations D.M. bridge trestle
Face, backfill and pave D.M. platform
Erect trestle and stairs D.M. platform
Harden foundations D.M. bridge trestle
Harden foundations U.L. bridge trestle
Break into U.L. platform
Foundations U.L. bridge trestle
Face, backfill and pave U.L. platform
Erect trestle and stairs U.L. platform
Break into D.L., U.M. platform and barrow off spoil
Harden foundations D.L., U.M. bridge trestle
Demolish old wall AD
Lay main tracks
Connect main tracks
Foundations D.L., U.M. bridge trestle
Face, backfill and pave D.L., U.M. platform
Erect trestle and stairs D.L., U.M. platform
Erect bridge over tracks and to N. wall
Lay local tracks
Connect local tracks
Lay bedstone in N. wall
Harden bedstone
Remove temporary bridge
Build 70yds S. wall EF
Build 30yds S. wall EF
Backfill (30+50) yds S. wall EF
Backfill 20yds S. wall EF
Build 30yds North wall DB
Build 50yds North wall DB
Backfill (30+30) yds North wall DB
Backfill 20yds North wall DB

Figure 8.3

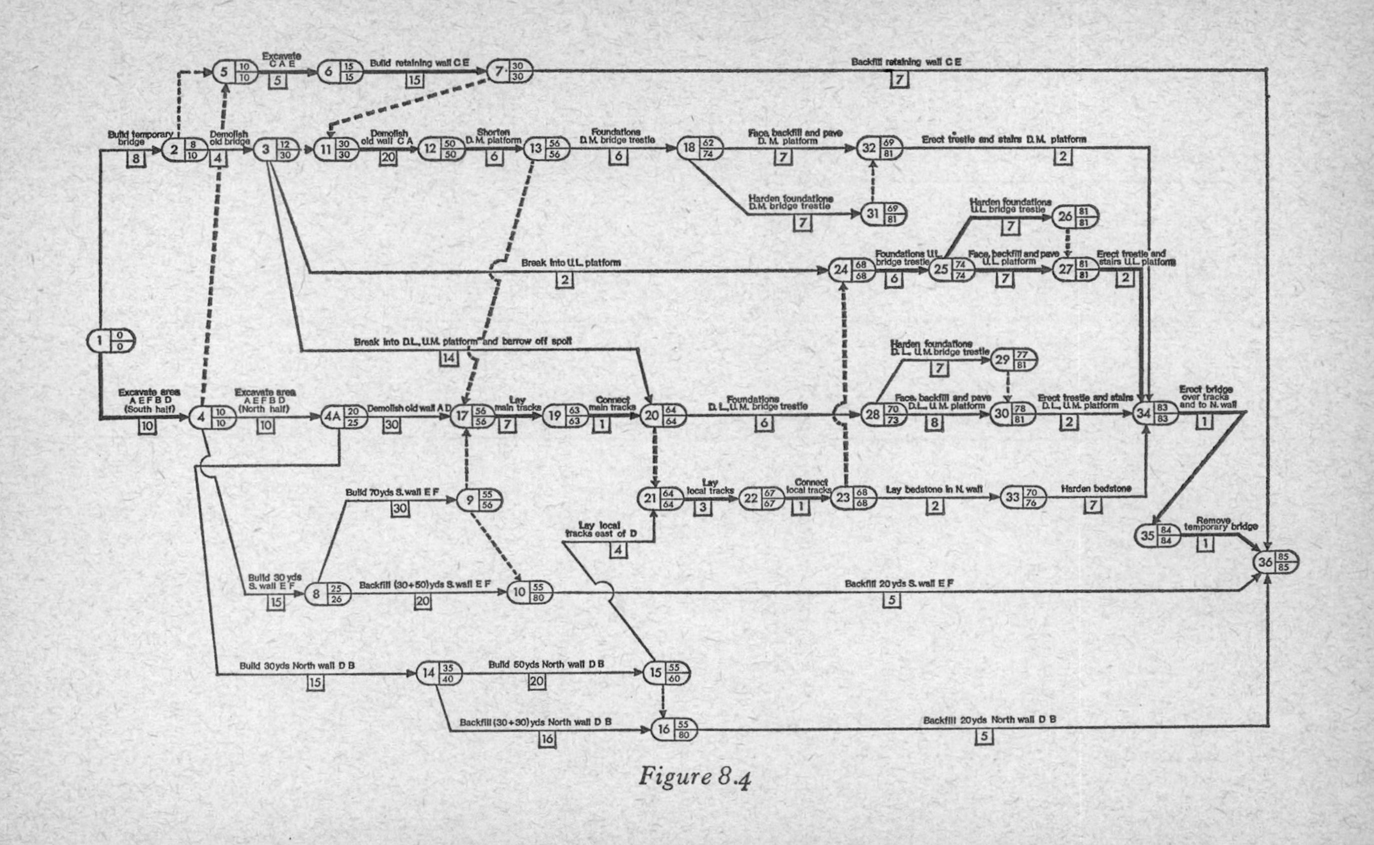
Build temporary bridge
Demolish old bridge
Excavate C A E
Build retaining wall C E
Backfill retaining wall C E
Demolish old wall C A
Shorten D.M. platform
Foundations D.M. bridge trestle
Face, backfill and pave D.M. platform
Erect trestle and stairs D.M. platform
Harden foundations D.M. bridge trestle
Harden foundations U.L. bridge trestle
Break into U.L. platform
Foundations U.L. bridge trestle
Face, backfill and pave U.L. platform
Erect trestle and stairs U.L. platform
Break into D.L., U.M. platform and barrow off spoil
Harden foundations D.L., U.M. bridge trestle
Excavate area A E F B D (South half)
Excavate area A E F B D (North half)
Demolish old wall A D
Lay main tracks
Connect main tracks
Foundations D.L., U.M. bridge trestle
Face, backfill and pave D.L., U.M. platform
Erect trestle and stairs D.L., U.M. platform
Erect bridge over tracks and to N. wall
Build 70yds S. wall E F
Lay local tracks
Connect local tracks
Lay bedstone in N. wall
Harden bedstone
Lay local tracks east of D
Remove temporary bridge
Build 30 yds S. wall E F
Backfill (30+50)yds S. wall E F
Backfill 20 yds S. wall E F
Build 30yds North wall D B
Build 50yds North wall D B
Backfill (30+30) yds North wall D B
Backfill 20yds North wall D B

Figure 8.4

construction of the north retaining wall (activity 14–15). A closer examination reveals that the major part of the local track can be laid from the eastward as far as the old retaining wall AD as soon as the north wall has been constructed without waiting for the removal of the old main tracks when the new main tracks are connected, and without waiting for the centre platform end to be demolished. The dummy 15–21 in the arrow diagram can be replaced by an activity of 4 days' duration representing laying that part of the local tracks east of D which need follow the construction of the north wall only, leaving in the critical path three days of laying local tracks following connecting main tracks and demolishing the centre platform end. This reduces the project duration by a further 4 working days.

The project length up to completion of the platforms is now reduced by 14 days to meet the Traffic Manager's objectives. The new plan is represented by the revised arrow diagram, Figure 8.4, page 82, in which it will be seen that the latest times for completion of all platforms (events 32, 27, and 30) are now 81 days as required, and the total length of the project is 85 working days. There is now of course no negative float.

The critical path is unchanged but the floats of many of the activities are reduced as shown in the revised table of activity times (Table 2, pages 84, 85).

The engineer notes in passing that some of the dummies have free float associated with them. Because event 29 has only dummies emanating from it (see page 64) the free float associated with dummy 29–30 could be made available to activity 28–29 'Harden foundations DL, UM bridge trestle' but in practice this is of no real value. The free float associated with the other dummies is meaningless.

A plan has now been formulated to meet the required objectives and it remains only for the Civil Engineer in charge to prepare a schedule and arrange for the necessary labour and machines to be available on site as required.

The schedule and the resources

Hitherto, during the planning stage, it has been possible to calculate times in numbers of working days from an arbitrary zero base line. In the preparation of the schedule the engineer must translate these numbers into calendar dates.

Table 2 **Alterations at Blandford Grove Station**

ANALYSIS OF NETWORK – REVISED

UL = Up Local DL = Down local UM = Up main
DM = Down main P = Possession (one day)

Activity description	Event Nos.	Duration	Earliest Start	Earliest Finish	Latest Start	Latest Finish	Total Float	Free Float
Build temporary bridge	1–2	8	0	8	2	10	2	0
Excavate area AEFBD Sth	1–4	10	0	10	0	10	0	0
Demolish old bridge	2–3	4	8	12	26	30	18	0
Dummy	2–5	0	8	8	10	10	2	2
Dummy	3–11	0	12	12	30	30	18	18
Break into DL, UM platform	3–20	14	12	26	50	64	38	38
Break into UL platform	3–24	2	12	14	66	68	54	54
Excavate area AEFBD Nth	4–4A	10	10	20	15	25	5	0
Dummy	4–5	0	10	10	10	10	0	0
Build 30 yds. S. wall EF	4–8	15	10	25	11	26	1	0
Build 30 yds. N. wall DB	4A–14	15	20	35	25	40	5	0
Demolish old wall AD	4A–17	30	20	50	26	56	6	6
Excavate area CAE	5–6	5	10	15	10	15	0	0
Build retaining wall CE	6–7	15	15	30	15	30	0	0
Dummy	7–11	0	30	30	30	30	0	0
Backfill retaining wall CE	7–36	7	30	37	78	85	48	48
Build 70 yds. wall EF	8–9	30	25	55	26	56	1	0
Backfill (30 + 50) yds. S. wall EF	8–10	20	25	45	60	80	35	10
Dummy	9–10	0	55	55	80	80	25	0
Dummy	9–17	0	55	55	56	56	1	1
Backfill 20 yds. S. wall EF	10–36	5	55	60	80	85	25	25
Demolish old wall CA	11–12	20	30	50	30	50	0	0
Shorten DM platform	12–13	6	50	56	50	56	0	0
Dummy	13–17	0	56	56	56	56	0	0
Foundations DM bridge trestle	13–18	6	56	62	68	74	12	0
Build 50 yds. N. wall DB	14–15	20	35	55	40	60	5	0
Backfill (30 + 30) yds. N. wall DB	14–16	16	35	51	64	80	29	4
Dummy	15–16	0	55	55	80	80	25	0
Lay local tracks east of D	15–21	4	55	59	60	64	5	5
Backfill 20 yds. N. wall DB	16–36	5	55	60	80	85	25	25
Lay main tracks	17–19	7	56	63	56	63	0	0

continued on next page

continued

Activity description	Event Nos.	Duration	Earliest Start	Earliest Finish	Latest Start	Latest Finish	Total Float	Free Float
Harden foundations DM bridge trestle	18–31	7	62	69	74	81	12	0
Face, backfill and pave DM platform	18–32	7	62	69	74	81	12	0
Connect main tracks	19–20	1	63	64	63	64	0	0
Dummy	20–21	0	64	64	64	64	0	0
Foundations DL, UM bridge trestle	20–28	6	64	70	67	73	3	0
Lay local tracks west of D	21–22	3	64	67	64	67	0	0
Connect local tracks	22–23	1	67	68	67	68	0	0
Dummy	23–24	0	68	68	68	68	0	0
Lay bedstone in N. wall	23–33	2	68	70	74	76	6	0
Foundations UL bridge trestle	24–25	6	68	74	68	74	0	0
Harden foundations UL bridge trestle	25–26	7	74	81	74	81	0	0
Face, backfill and pave UL platform	25–27	7	74	81	74	81	0	0
Dummy	26–27	0	81	81	81	81	0	0
Erect trestle and stairs UL platform	27–34	2	81	83	81	83	0	0
Harden foundations DL, UM bridge trestle	28–29	7	70	77	74	81	4	0
Face, backfill and pave DL, UM platform	28–30	8	70	78	73	81	3	0
Dummy	29–30	0	77	77	81	81	4	1*
Erect trestle and stairs DL, UM platform	30–34	2	78	80	81	83	3	3
Dummy	31–32	0	69	69	81	81	12	0
Erect trestle and stairs DM platform	32–34	2	69	71	81	83	12	12
Harden bedstone	33–34	7	70	77	76	83	6	6
Erect bridge over tracks and to N. wall	34–35	1	83	84	83	84	0	0
Remove temporary bridge	35–36	1	84	85	84	85	0	0

* This free float can theoretically be made available to 28–29 'Harden Foundations DL, UM Bridge Trestle' but this is meaningless in practice.

(I) 'POSSESSION' OF THE TRACK

The engineer in charge was told that access to the Blandford Grove site from the eastward would be available from 24th February 1964 and that the reconstruction of the platforms must be completed to allow double-stopping of trains to cease, before the summer time-table commenced on 15th June. In this period there are 77 working days assuming a five-day week and excluding Good Friday, Easter Monday and Whit Monday. In addition however it was agreed that work would have to take place on four Sundays when the traffic on the tracks could be suspended in order to:

1 build temporary bridge
2 demolish old bridge
3 connect main tracks
4 connect local tracks.

(A further two Sunday 'possessions' of the track were to be arranged at a later date for erecting the new bridge and removing the temporary bridge.)

In the period between 24th February and 14th June therefore there are 81 working days including the four Sunday 'possessions'.

The engineer now realizes that the 3rd and 4th of these 'possessions' are on the critical path and are required on the 64th and the 68th working days. Clearly they cannot both be on Sunday unless some special arrangement is made. Each of the other two 'possessions' is required on the last day of an activity with float and is therefore flexible.

Because the dates of these 'possessions' determine which days are working days and which are rest days, at least as far as the critical path is concerned, the engineer arranges with the Traffic Manager that traffic will be diverted round Blandford Grove on these days (see Figure 8.5, page 87):

1 8th March—to erect temporary bridge
2 15th March—to demolish old bridge
3 24th March—to connect main tracks
4 31st May—to connect local tracks.

This means that there will be a delay of one day in the critical path prior to 24th May, and two days' delay between 24th and 31st May. If one of these delay days falls between the EST and LFT of an activity, that activity in effect gains one extra day's float.

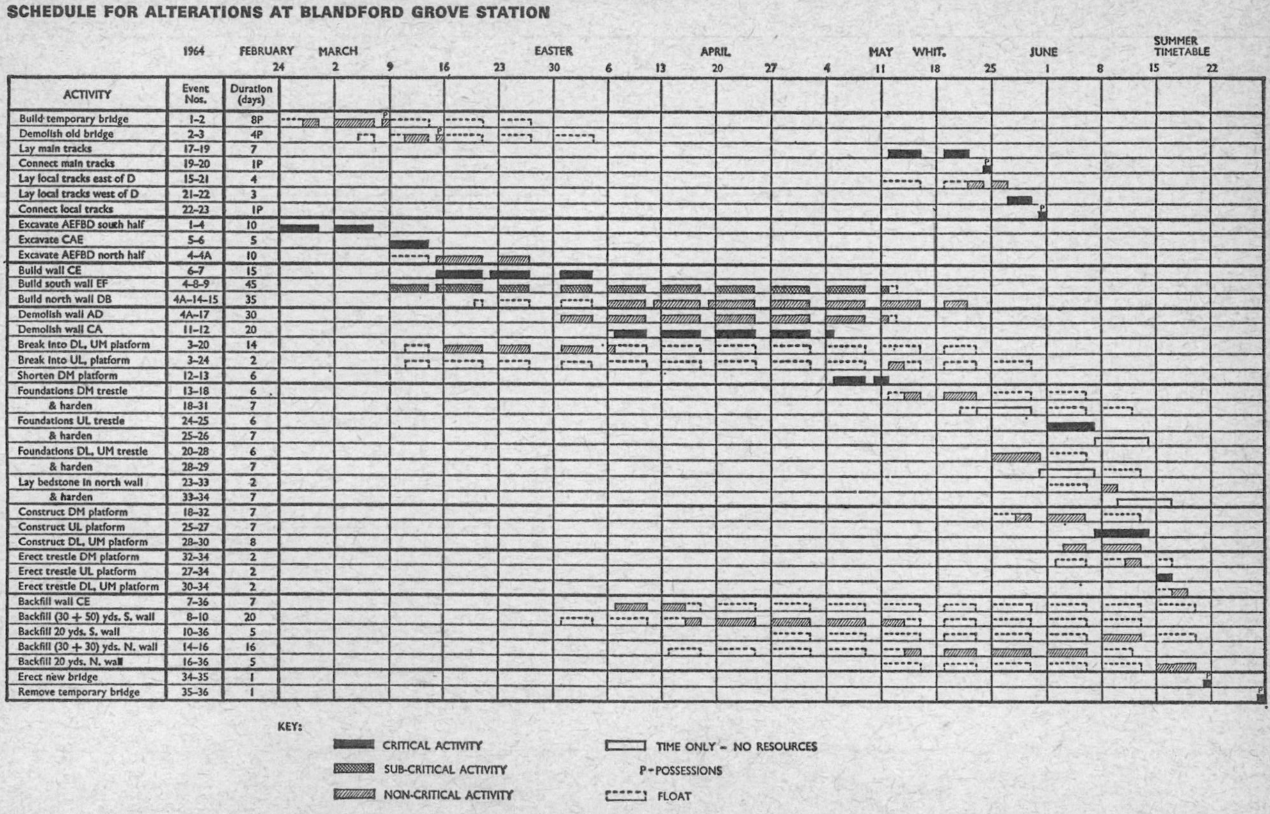

SCHEDULE FOR ALTERATIONS AT BLANDFORD GROVE STATION

1964 — FEBRUARY 24 — MARCH 2, 9, 16, 23 — EASTER 30 — APRIL 6, 13, 20, 27 — MAY 4, 11 — WHIT. 18, 25 — JUNE 1, 8, 15 — SUMMER TIMETABLE 22

ACTIVITY	Event Nos.	Duration (days)
Build temporary bridge	1–2	8P
Demolish old bridge	2–3	4P
Lay main tracks	17–19	7
Connect main tracks	19–20	1P
Lay local tracks east of D	15–21	4
Lay local tracks west of D	21–22	3
Connect local tracks	22–23	1P
Excavate AEFBD south half	1–4	10
Excavate CAE	5–6	5
Excavate AEFBD north half	4–4A	10
Build wall CE	6–7	15
Build south wall EF	4–8–9	45
Build north wall DB	4A–14–15	35
Demolish wall AD	4A–17	30
Demolish wall CA	11–12	20
Break into DL, UM platform	3–20	14
Break into UL, platform	3–24	2
Shorten DM platform	12–13	6
Foundations DM trestle	13–18	6
& harden	18–31	7
Foundations UL trestle	24–25	6
& harden	25–26	7
Foundations DL, UM trestle	20–28	6
& harden	28–29	7
Lay bedstone in north wall	23–33	2
& harden	33–34	7
Construct DM platform	18–32	7
Construct UL platform	25–27	7
Construct DL, UM platform	28–30	8
Erect trestle DM platform	32–34	2
Erect trestle UL platform	27–34	2
Erect trestle DL, UM platform	30–34	2
Backfill wall CE	7–36	7
Backfill (30 + 50) yds. S. wall	8–10	20
Backfill 20 yds. S. wall	10–36	5
Backfill (30 + 30) yds. N. wall	14–16	16
Backfill 20 yds. N. wall	16–36	5
Erect new bridge	34–35	1
Remove temporary bridge	35–36	1

KEY:

- CRITICAL ACTIVITY
- SUB-CRITICAL ACTIVITY
- NON-CRITICAL ACTIVITY
- TIME ONLY – NO RESOURCES
- P–POSSESSIONS
- FLOAT

Figure 8.5

The engineer realizes these delays are unavoidable and arranges to make up the time by overtime working on critical activities on Saturday 6th, Sunday 7th and Saturday 13th June, leaving Sunday 14th spare for coping with any last-minute emergencies.

This fixes the 81 working days for the critical activities but does not preclude weekend overtime working on critical activities either because an estimated duration has proved in error, or to help smooth the demand for resources. Neither does it commit the engineer to overtime working on non-critical activities on 'possession' days if within the limits of total float available such overtime can be avoided.

(II) EXCAVATION OPERATIONS

Excavation of the south half of area AEFBD is the only activity which the engineer arranges to start on 24th February. By taking advantage of the float available in excavating the north half he can arrange for the three excavation operations to be carried out consecutively by the same team, as shown in the bar chart in Figure 8.5, page 87. Because there are two 'possessions' on 8th and 15th March, and only one lost day prior to the 'possession' on 24th May, the engineer decides it is necessary to work overtime on excavation on either 8th or 15th March to complete all excavations in effect in 25 working days which represents the limit of the float available.

(III) BUILDING WALLS

Even taking maximum advantage of the float associated with the construction of the three walls it appears at first sight that for some five days three gangs of bricklayers will have to be on the site simultaneously. The engineer decides it is more economical to authorize three Sundays of overtime so that one gang can build retaining wall CE and the north wall DB as consecutive operations while a second gang builds the south wall.

(IV) BACKFILLING BEHIND WALLS

Backfilling operations are arranged consecutively starting at their latest starting times to ensure that they will not overtake wall construction. The engineer is satisfied that all the backfilling operations can be carried out consecutively by one gang within the limits of float available.

(v) DEMOLITION OF WALLS AND PLATFORM ENDS

There is insufficient float available to prevent the demolition of walls AD and CA being carried out concurrently. Because of the one-day delay introduced by arranging the third 'possession' for 24th May however it is possible to delay the critical demolition of wall CA by one day leaving 14 days for breaking into the centre platform prior to demolishing wall CA. The same gang which carries out these two operations can follow on with the demolition of the down-main platform then the up-local platform, so that no more than two demolition gangs are required at any one time. The engineer makes arrangements accordingly.

(vi) CONSTRUCTION OF NEW PLATFORMS

There is insufficient float to enable all three new platforms to be constructed by the same gang of men, but the engineer arranges a small gang to construct the down-main and up-local platforms consecutively and a larger gang to construct the centre platform.

(vii) FOUNDATIONS AND BEDSTONE

The engineer again decides it is worth while to authorize overtime on two days at weekends to enable all the bridge trestle foundations and the bedstone in the north wall to be carried out as consecutive operations by one gang of men.

(viii) ERECTING BRIDGE TRESTLES

The engineer again authorizes steel erectors to work overtime through the last weekend before the summer timetable starts to enable them to erect the three bridge trestles consecutively.

The schedule in operation

The engineer in charge has now formulated his schedule of work. He has in many cases taken advantage of the float available to enable him to schedule a particular operation on a particular scheduled 'possession' Sunday, or to smooth out possible peaks of demand for resources, or to guide him in the authorization of overtime.

The resulting schedule (Figure 8.5, page 87) is practical and strikes a reasonable balance between, on the one hand, the desire to take maximum advantage of the available float to help smooth demand for resources, and on the other hand, the dangers of ending up with a schedule which is too rigid. Flexibility in the form of further weekend or weekday overtime remains should there be a need to counteract delays due to exceptionally bad weather, strikes or other unforeseeable happenings.

The bar chart provides the foreman with a sound basis for carrying out the project and it provides the manager with a means of supervising progress. The ease with which the arrow diagram and schedule can be updated to assess the effects of changing circumstances enables the manager to act promptly, forestall trouble, and virtually guarantee the completion date.

Nine Conclusion

The emergence and development of Network Planning Techniques in their present form have been encouraged and stimulated in recent years by the need for better means of planning and controlling projects of ever-increasing size and complexity. There is however nothing fundamentally new in these techniques. They only provide in fact a visual and numerical means of doing what must always be done in embarking on a project, namely fitting the constituent parts of the project together in their logical sequences and assessing the implications of these sequences on scheduling.

These techniques do however enable much larger and more complex projects to be planned in much greater detail and with much greater precision than has perhaps been possible with the use of traditional planning systems. In so doing they not only provide a means of setting up an initial plan for the project in hand, they also provide a system for co-ordinating the many activities in the project, monitoring their progress against realistic target dates for each, and assessing well in advance the effects of unforeseen contingencies on the project plan as a whole so that the necessary action can be taken, and the necessary degree of control can be maintained.

The technique will not however of itself manage the project, nor in any way relieve management of the responsibility for making decisions. But with the use of Network Planning Techniques management has much more reliable information on which to base its decisions, and is in a position better to assess the value to the project as a whole of alternative courses of action.

Although the principles of the technique as presented in this publication are simple enough their application in practice is not without problems. Preparing the arrow diagram involves in a sense living through the project in advance. Decisions have to be made at this stage as to the methods and equipment to be employed. These are decisions which would however have to be made eventually in any case and the fact that they are taken early does not make them irrevocable. The arrow diagram incorporating these decisions yields data which are valid only so long as these decisions remain unaltered.

The identification of the true sequences and the best way to represent them in the arrow diagram are ever-present problems despite the apparent simplicity of asking 'what must precede ...?' and 'what must follow ...?' Endeavouring to show all the interdependencies with an uncompromising exactness tends to lead to more detail than is really required. For example, erecting a single lamp-post must clearly precede wiring it to the electricity supply. Erecting and wiring a series of lamp-posts on a housing estate however might be shown as two overlapping activities, but because the stagger by which the erector leads the electrician is so small, perhaps less than a day, it may be sufficient to show erecting and wiring as a single combined activity : 'erect and wire lamp-posts'.

The boundaries of the project itself are not always clearly defined. Should the arrow diagram be drawn for a single house, or for an estate of houses including the installation of services? Everything that is added makes the project larger and more unwieldy, anything that is left out means a risk of overlooking some vital interdependency which might alter the whole situation. The only solution is to consider the situation on the broadest possible scale at the outset to make sure that everything which might conceivably affect the project has been taken into account. It may then be possible to exclude from the initial arrow diagram those parts of the work which clearly have no vital influence on the project completion date.

Arrow diagrams should be prepared quickly. Undue preoccupation with detail in the early stages may turn out to be effort wasted on non-critical parts of the project. The arrow diagram is admittedly only as good as the information built into it but refinements can come later. The sooner it can be prepared and analysed even in an imperfect form, the sooner the arrow diagram will fulfil its primary function of directing attention to the critical areas of the project.

It is this feature of the technique, and the ability to update quickly and bring to light shifts in the critical areas of the project, which enables management to co-ordinate and control very complex projects with greater ease and efficiency. This alone makes Network Planning Techniques an addition of immense value to the body of management techniques.

Appendix One The two original network planning systems—P E R T and C P M

The foregoing chapters have dealt solely with the principles and mechanics of network planning. These principles underlie all the various network planning systems, mainly computer-based, which have appeared in recent years, for example, PEP (Programmed Evaluation Procedure), RPSM (Resources Planning and Scheduling Method), LCS (Least Cost Scheduling), MOSS (Multi-operation Scheduling System) etc.

Not only because they were first in the field but also because of the spectacular results they achieved, the two original systems, namely PERT (Programme Evaluation and Review Technique) and CPM (Critical Path Method) have received a great deal of attention in the technical press. It is felt therefore that this booklet would not be complete without a short exposition of the principles of these two special systems.

Section 1 PERT

When the US Navy was faced in 1957 with meeting a target date for the completion of the Polaris missile project two years in advance of the date originally predicted, it was decided that traditional planning methods would not give adequate control of the many activities and contracts involved in the project. Together with Booz, Allen, and Hamilton, a firm of consultants, a new system of planning known as Programme Evaluation and Review Technique was devised.

The essential feature of the Polaris project was that much of the work to be done was research and development work and therefore not easy to define either in terms of time required or resources required. The events were however readily definable. In other words it was much easier to say that component *X* had to be completed and tested before sub-assembly *Y* could be built, than to define with any degree of certainty the time required or the work involved in completing and testing component *X* or building sub-assembly *Y*.

It was in this essentially uncertain situation that the PERT system was born. In order to deal with these uncertainties some basic assumptions were made as to the statistical probabilities of completing each activity in a particular time. These assumptions have been called into question by other authorities and it is necessary to bear this in mind when considering the principles and practical uses of PERT.

Basic concept

The basic concept of PERT is to calculate the probabilities of reaching particular events in the network by specified target dates, having regard to the possible variations in the activity times.

The system is predominantly concerned with events and is therefore an event-orientated system.

Basic parameters

In order to deal with the uncertainties associated with the activities, the PERT system requires three estimates of duration to be prepared for each activity :

(i) t_p= the pessimistic time
i.e. the time the activity would take if every possible delay occurred
(ii) t^o = the optimistic time
i.e. the best time that could possibly be achieved if the activity is completed without a hitch
(iii) t_m = the most like time
i.e. the time the activity would normally be expected to take.

Basic assumptions

The fundamental assumption in PERT, in its most simplified form, was that if a particular activity was carried out on a large number of occasions the actual times taken would form a frequency distribution very similar to the Beta distribution with t_o and t_p at the tails and t_m at the peak frequency (Figure A.1).

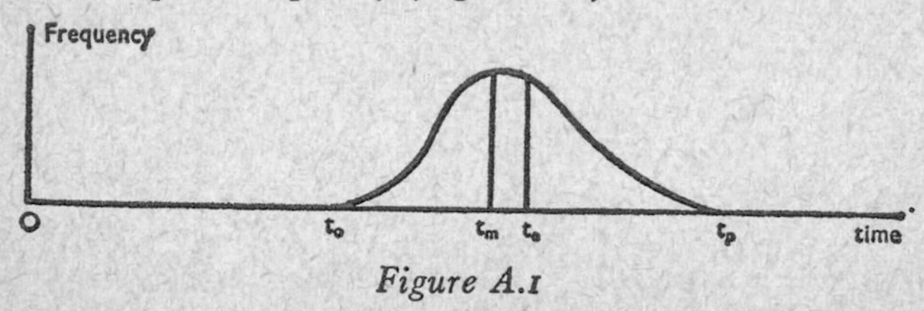

Figure A.1

This being so it was further assumed that a fair approximation of the average or expected time (t_e) and of the standard deviation (σ_{t_e}) of the distribution were given by:

$$t_e = \frac{t_o + 4t_m + t_p}{6}$$

$$\text{and } \sigma_{t_e} = \frac{t_p - t_o}{6} \quad (\text{Variance} = \sigma^2_{t_e})$$

Some authorities have subsequently questioned the validity of these assumptions. For instance it has been suggested that the Gamma distribution would be more suitable. It has also been pointed out that although the above formula for t_e gives a fair approximation of the average of the Beta distribution in most circumstances, the standard deviation calculated from one-sixth of the range often tends to give an estimate lower than the true value. This should be borne in mind when calculating probabilities based on these standard deviations.

Event times

The earliest and latest times, T_E and T_L, for the events in the network, are calculated in the normal way as explained in Chapter 5, using the calculated expected times, t_e, for activity durations. In what follows only T_E is mentioned but similar arguments and deductions apply also in connection with T_L.

An event earliest time calculated in this way is clearly only one of a distribution of many possible event times arising from various combinations of the possible activity times up to the event. If the network is large enough for statistical principles to apply, this distribution will be an approximately normal (Gaussian) distribution becoming more nearly normal as the size of the network increases.

Since variances are additive, the variance ($\sigma^2_{T_E}$) of the normal distribution associated with an event earliest time, T_E, is the sum of the variances of activities in the chain of activities controlling that event time. This is not the largest sum of variances up to that event, but the sum of variances on the longest path from the start of the project up to the event. These variances, $\sigma^2_{T_E}$, can be readily calculated and the standard deviation, σ_{T_E}, of the distribution of T_E is then the square root of the variance.

Probability of meeting target date

Having calculated the standard deviation, σ_{T_E} associated with each event earliest time it is then possible to calculate the probability of reaching any particular event in the network by any specified target date, T_S. This probability is equivalent to the area under the normal distribution curve and to the left of the vertical representing T_S, in other words the shaded area in Figure A.2. If the ratio $\frac{T_S - T_E}{\sigma_{T_E}}$ is calculated the equivalent shaded area can be obtained by reference to statistical tables of areas under the normal curve. This area is virtually zero when $T_S = T_E - 3\sigma_{T_E}$, 0·5 when $T_S = T_E$ and nearly 1·0 when $T_S = T_E + 3\sigma_{T_E}$, representing probabilities of zero, 50 per cent and 100 per cent respectively.

In practice an event target date with a probability of less than

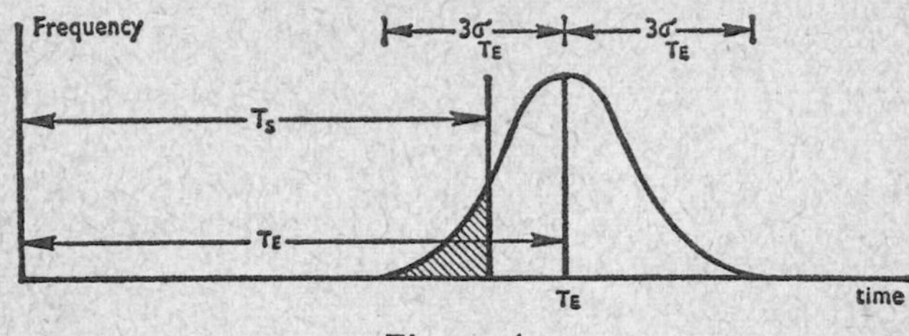

Figure A.2

0·4 is unlikely to be achieved and more resources should be allocated to the chain of activities controlling that event time. A probability greater than 0·6 is so likely to be achieved that it would be possible to consider reducing the resources in the controlling path of activities up to the event. If the probability is between 0·4 and 0·6 there is a satisfactory chance of reaching the event by the specified target date with the planned resources.

It should be emphasized however that probabilities calculated in this way assume an approximately normal distribution for event times, which is the case in practice only if the network is large.

Uses of PERT

The PERT system was originally devised to deal with a particular situation, namely the situation in which the durations of the activities in the project were uncertain and unpredictable. This condition normally obtains in research and development work, indeed in any kind of investigation work, and to some extent in design work. For planning projects in these spheres therefore the use of PERT in its original form may be of value provided the network is large. For the 'run-of-the-mill' construction or maintenance project previous experience enables reasonably accurate estimates of durations to be prepared for most activities. There is therefore little to be gained in employing the complicated PERT system for this type of project—a simple one-estimate system is generally adequate.

Because of the large number of calculations involved, PERT is essentially a computer-based system. Several computer firms advertise PERT programmes for hire, which are similar to the original PERT programme. They can of course be used to carry out straightforward network analyses by bypassing those parts of the programme which calculate expected times and probabilities.

Section 2 **CPM**

The cost in loss of revenue of taking multi-million pound chemical plants out of commission at regular intervals for maintenance and overhaul prompted the Du Pont chemical company in America to seek ways of reducing off-line time. One line of attack was to reduce the time required for an overhaul by careful planning of the work in advance. A network planning system known as Critical Path Method was devised for this purpose, quite independently of the concurrent development of PERT for the Polaris project. CPM was largely the brainchild of Walker of Du Pont, working with Kelley the mathematician and Mauchly of Remington Rand, who provided the computer knowhow.

In contrast to the inherent uncertainties surrounding the activities of the Polaris project, the times required for the tasks or activities in overhauling a chemical plant are reasonably amenable to estimation within acceptable limits of accuracy by the use of Work Study and other appropriate techniques. This applies also to times required to accomplish these tasks by various methods and under various conditions and to the direct costs associated with each.

Basic concept

The basic concept of CPM is to calculate the project duration which will result in the minimum over-all project cost assuming that there is an approximately linear and measurable relationship between the time and direct cost of each activity.

The system is predominantly concerned with activities and is therefore an activity-orientated system.

Basic parameters

The CPM system requires for each activity the preparation of four values:

(i) d_n = normal duration
i.e. the time the activity would take if carried out by the most economical method under 'normal' conditions
(ii) c_n = normal cost
i.e. the direct cost associated with duration d_n
(iii) d_c = crash duration
i.e. the minimum duration it is possible to achieve
(iv) c_c = crash cost
i.e. the direct cost associated with duration d_c.

Since c_n is the cost for carrying out the activity by the most economical method, the crash cost will be greater than the normal cost, the excess representing the cost of overtime, hiring extra

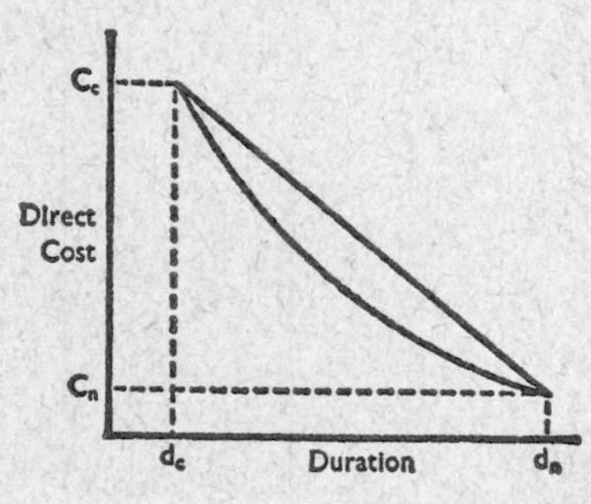

Figure A.3

machines, labour, and so on. In general the direct cost of an activity will rise as the duration falls but the relationship between direct cost and duration is unlikely to be linear. Normally it will be a curve of the form in Figure A.3.

At the expense of a little accuracy however the relationship may be assumed to be linear as represented by the straight line joining the normal and crash points in the diagram. The slope of this line represents the extra cost incurred per unit of time saved on the activity duration.

$$\text{i.e. extra cost per unit time} = \frac{c_c - c_n}{d_n - d_c}$$

Expediting the project at minimum direct cost

It is assumed that the network has been set up and analysed using the normal durations for all the activities. The completion date in these circumstances is therefore known.

There are usually several ways of expediting a complex project. Clearly each of the critical and near critical activities must be examined for possible shortening and for the unit cost of doing so. Various combinations of reductions of activity durations could result in reducing the project duration by a specified amount of time, say one time unit. The cost of each of these must be evaluated and the minimum selected.

This process can be continued, reducing the project duration by steps of one time unit, calculating for each project duration the minimum direct project cost, until the minimum (i.e. crash) project duration is reached. The project direct cost and duration will be related as shown in Figure A.4.

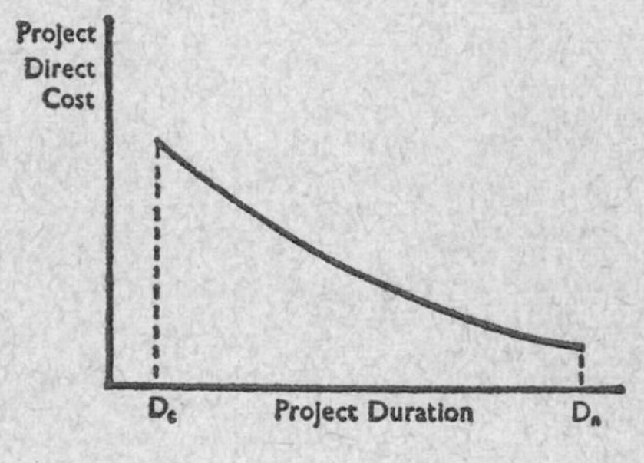

Figure A.4

Indirect project cost

The indirect costs associated with each of a range of project durations must next be evaluated. What to include in indirect costs is not always easy to decide but certainly any costs directly related to the length of the project must be regarded as indirect costs. These will comprise such things as the cost of loss of revenue, the cost of administrative overheads and the cost of retention of machines and

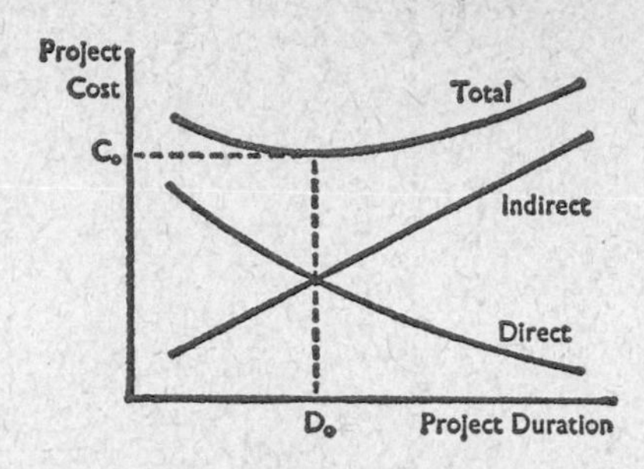

Figure A.5

equipment of a general nature not directly chargeable to specific activities, e.g. compressors, scaffolding etc. Also included on the positive and negative side respectively will be the cost of any penalty or bonus arising in connection with particular project completion times.

In general the indirect cost will increase as the project duration increases though the relationship will not necessarily be linear.

Over-all project cost

If the total of direct cost and indirect cost is plotted on the graph against the project duration, a minimum cost, C_o, will be found, with an associated duration, D_o. There is nothing to be gained by reducing the project duration below D_o because this is the duration for which the over-all project cost is a minimum.

Uses of CPM

CPM is beginning to be used in the construction industry where there is a vast body of knowledge and experience to enable accurate estimates to be prepared. It has limitations of course. Apart from the difficulty in establishing valid cost/time relationships for the activities, and the error introduced by the assumption that the relationship between activity cost and duration is linear, there may be little return for the effort involved in using a relatively elaborate system in a project where indirect cost so far outweighs direct cost as to make minimization of project time the main objective.

Because of the many calculations involved CPM, like PERT, is essentially a computer-based system. CPM programmes can however be used for straightforward network analysis based on the provision of normal times only and bypassing those parts of the programme which deal with the cost calculations.

Section 3 Conclusion

Whereas PERT has its applications in those projects in which there is some uncertainty about the activities but the necessary resources, including money, are always available as required, CPM has its applications in projects where the activities are definable and measurable and minimum over-all cost is of the utmost importance.

This does not mean however that the use of PERT will be confined to government defence projects like the Polaris project, nor that CPM will be of interest only to profit-making concerns. The merits of each system, and indeed of systems developed from these and now available, should always be considered in relation to the project in hand.

Appendix Two Examples of computer input forms and printed output forms

WASHMACO Ltd. Form 1234/IP/D1 — **ICL 1900 Series PERT Activity Cards S**

Standard Network Module D — Input Sheet Number 1 — Project DR — Subproject 66 — Unit CABINET

Class	Project code	Sub-project code	Event name or prec event	Succ. event	Identifier	Time	Type	Description	Report code / Resp. cost	Date	Date type	Resources (Res. cat. / Quantity)	des. S	Card num.
A	DR	66	201	210		001		PRINT DRAWINGS	DO				S	
A			01	14				PRINT DRAWINGS	DO	note deleted activity			S	
A	DR	66	201	202		001		PRINT DRAWINGS	DO				S	
A	DR	66	202	203		022	L	PLAN PIECE PARTS	PLN				S	
A	DR	66	202	204		002		PLAN FIRST PIECE PARTS	PLN				S	
A	DR	66	202	208		002		PLAN FIRST PIECE PARTS	PLN				S	
A	DR	66	204	205		010	L	DESIGN TOOLS	TDM				S	
A	DR	66	204	206		010		DESIGN FIRST TOOLS	TDM				S	
A	DR	66	203	205		003		DESIGN LAST TOOLS	TDM				S	
A	DR	66	206	207		040	L	MANUFACTURE TOOLS	TDM				S	
A	DR	66	206	217		020		MANUFACTURE FIRST TOOLS	TDM				S	
A	DR	66	205	207		020		MANUFACTURE LAST TOOLS	TDM				S	
A	DR	66	208	209		012	L	PREPARE PARTS DOCUMENTS	PC				S	
A	DR	66	208	217		001		PREP FIRST PARTS DOCS	PC				S	
A	DR	66	203	209		001		PREP LAST PARTS DOCS	PC				S	
A			14	15				PREP LIST PURCHASE ITEMS	PC	note deleted activities			S	
A			15	16				ORDER OUTSIDE PURCHASES	DO				S	
A			16	19				OBTAIN OUTSIDE PURCHASES	DO				S	
A	DR	66	210	211		010	L	PLAN ASSEMBLIES	PLN				S	
[illegible]	[illegible]	66	210	212		[illegible]		[illegible]ST ASSEMBLIES	[illegible]				S	
[illegible]			2[illegible]										S	
A	D[illegible]	[illegible]	207	218		010		MANUFACTU[illegible] [illegible]ARTS	PRD					
A	DR	66	209	218		010		MANUFACTURE LAST PARTS	PRD					
A	DR	66	219	220		010	L	MANUFACTURE ASSEMBLIES	PRD				S	
A	DR	66	218	220		003		MANUFACTURE LAST ASSYS	PRD				S	
A	DR	66	213	220		003		MANUFACTURE LAST ASSYS	PRD				S	
A	DR	66	220	221		004		ASSEMBLE UNIT	PRD				S	

Figure A2.1 Preprinted standard input data sheet

I C L 1900 SERIES PERT 21/11/69 OUTPUT SHEET NUMBER 22

PROJECT DB TEST NETWORK - PST 2.3 RUN TIME NOW 5DEC66 PAGE 1

TIME ANALYSIS IN TOTAL FLOAT AND EARLIEST START SEQUENCE

S/P CDE	PREC EVENT	SUCC EVENT	REPORT CODE	DESCRIPTION	DUR	EARLIEST START	EARLIEST FINISH	LATEST START	LATEST FINISH	TOT FLOAT	FREE E FLT
P3	1	2	TEC	DESIGN	3.0	5DEC66T	22DEC66	2DEC66	21DEC66	-.1	.0
P3	2	4	TEC	PLAN CASTINGS	2.0	22DEC66	9JAN67	21DEC66	6JAN67	-.1	.0
P3	4	14	PUR	OBTAIN CASTINGS	12.0	9JAN67	3APR67	6JAN67	31MAR67	-.1	.0
P3	14	15	PRD	M/C CASTINGS	4.0	3APR67	1MAY67	31MAR67	28APR67	-.1	.0
P3	15	16	QAD	INSPECT	1.0	1MAY67	8MAY67	28APR67	5MAY67	-.1	.0
P3	16	21	PRD	MECH ASSEM	2.0	8MAY67	22MAY67	5MAY67	19MAY67	-.1	.0
P3	21	29	PRD	RUN-IN	1.0	22MAY67	30MAY67	19MAY67	26MAY67	-.1	.0
P3	29	30	PRD	FIT CONTROLS	1.0	30MAY67	6JUN67	26MAY67	5JUN67	-.1	.0
P3	30	31	QAD	TEST	2.0	6JUN67	20JUN67	5JUN67	19JUN67	-.1	.0
[illegible]	1	2	TEC	DESIGN	5.0	5DEC66T	9JAN67	5DEC66T	9JAN67	.0	[illegible]
	2	3	TEC	PLANNING	3.0	9JAN67	30JAN67	9JAN67	10MAR67	.0	
		[illegible]	TEC	DESIGN BASE FRAME						[illegible]	
P1	23				1.0	27JUN67	4JUL67	27JUN67	[illegible]	[illegible]	.0
P1	24	[illegible]		ACCEPTANCE TEST	2.0	4JUL67	18JUL67	4JUL67	18JUL67	.0	.0
P1	2	6		LEAD	1.0	9JAN67	16JAN67	13JAN67	20JAN67	.4	.0
P1 L	6	7	TEC	TOOL DESIGN	6.0	16JAN67	27FEB67	20JAN67	22MAR67	.4	.0
P1	6	8		LEAD	2.0	16JAN67	30JAN67	20JAN67	3FEB67	.4	.0
P1 L	8	9	PRD	TOOL MANUFACTURE	10.0	30JAN67	10APR67	3FEB67	14APR67	.4	.0
P1 L	10	11	PRD	P/P MANU	8.0	20FEB67	24APR67	3MAR67	28APR67	.4	.0
P1	9	11		LAG	2.0	10APR67	24APR67	14APR67	28APR67	.4	.0
P1	11	12	PRD	SUB-ASSEM	2.0	24APR67	8MAY67	28APR67	12MAY67	.4	.0
P1	12	18	PRD	STAGE 1 ASSY	1.0	8MAY67	15MAY67	12MAY67	19MAY67	.4	[illegible]
P1					[illegible]	[illegible]		[illegible]	26MAY67	.4	
				LEAD							.0
P1	2	14	TEC	DESIGN CIRCUITS	3.0			3FEB67	24FEB67	3.4	.0
P1 L	4	5	PUR	OBTAIN RAW MAT'L	9.0	16JAN67	18MAR67	10FEB67	14APR67	3.4	.0
P1	4	10		LEAD	3.0	16JAN67	6FEB67	10FEB67	3MAR67	3.4	2.0
P1	14	15 A	PUR	OBTAIN B.O.ITEMS	11.0	30JAN67	17APR67	24FEB67	12MAY67	3.4	1.0
P1	7	9		LAG	3.0	27FEB67	18MAR67	22MAR67	14APR67	3.4	3.0
P1	5	11		LAG	2.0	18MAR67	3APR67	14APR67	28APR67	3.4	3.0
P1	13	12	PRD	MAKE BASE FRAMES	4.0	18MAR67	17APR67	14APR67	12MAY67	3.4	3.0
P1	18	22	PRD	ST. 1 ASSY	1.0	15MAY67	22MAY67	13JUN67	20JUN67	4.0	.0
P1	22	23	PRD	STAGE 2 ASSEM	1.0	22MAY67	30MAY67	20JUN67	27JUN67	4.0	4.0
P3	2	5	TEC	PLAN COMPONENTS	3.0	22DEC66	16JAN67	27JAN67	17FEB67	4.4	.0
P3	5	7	PRD	MAKE COMPONENTS	6.0	16JAN67	27FEB67	17FEB67	31MAR67	4.4	.0

Figure A2.2 Time analysis report in total float and earliest start sequence

```
I C L  1900 SERIES PERT                                  22/10/69                          OUTPUT SHEET NUMBER   32

                         PROJECT  DB   TEST NETWORK - PST 2.3                    RUN    1  TIME NOW  5DEC66  PAGE   1

 TIME ANALYSIS BAR CHART IN REPORT CODE AND EARLIEST START SEQUENCE

                                                            5         19       2       16       30       13       27
S/P  PREC   SUCC U  REPORT    DESCRIPTION        DUR   DEC66       DEC66   JAN67    JAN67    JAN67    FEB67    FEB67
     EVENT  EVENT I  CODE                                X....*......I....*..I....*....I....*....I....*....I....*....I...
P1      2      6              LEAD               1.0     X    *      I    *  I   AAAAA.... *    I    *    I    *    I
P1      2      4              LEAD               1.0     X    *      I    *  I   AAAAA...................I    *    I
P1      6      8              LEAD               2.0     X    *      I    *  I   *    AAAAAAAAAA.... *    I    *    I
P1      4     10              LEAD               3.0     X    *      I    *  I   *    AAAAAAAAAAAAAAA.................
P1      8     10              LEAD               3.0     X    *      I    *  I   *    I    *    AAAAAAAAAAAAAAA.........
P1      3      7              LAG                2.0     X    *      I    *  I   *    I    *    AAAAAAAAAA.............+
P1      3      5              LAG                4.0     X    *      I    *  I   *    I    *    AAAAAAAAAAAAAAAAAAAA....+
P1      7      9              LAG                3.0     X    *      I    *  I   *    I    *    I    *    I    *    AAAA+
P1      5     11              LAG                2.0     X    *      I    *  I   *    I    *    I    *    I    *    I   +
P1      9     11              LAG                2.0     X    *      I    *  I   *    I    *    I    *    I    *    I   +
P1     11     15              DUMMY               .0     X    *      I    *  I   *    I    *    I    *    I    *    I   +
P1     19     23              DUMMY               .0     X    *      I    *  I   *    I    *    I    *    I    *    I   +
P1      8      9    PRD       TOOL MANUFACTURE  10.0     X    *      I    *  I   *    I    *    AAAAAAAAAAAAAAAAAAAAAAAAA+
L                                                        X    *      I    *  I   *    I    *    I    *    I    *    I
P1     14     15 B  PRD       MAKE SPECIAL       5.0     X    *      I    *  I   *    I    *    AAAAAAAAAAAAAAAAAAAAAAAAA+
                              ITEMS               .      X    *      I    *  I   *    I    *    I    *    I    *    I
P1     10     11    PRD       P/P MANU           8.0     X    *      I    *  I   *    I    *    I    *    I    AAAAAAAAA+
L                                                        X    *      I    *  I   *    I    *    I    *    I    *    I
P1     13     12    PRD       MAKE BASE FRAMES   4.0     X    *      I    *  I   *    I    *    I    *    I    *    I   +
P1     16     17    PRD       MAKE COVERS M/C1   6.0     X    *      I    *  I   *    I    *    I    *    I    *    I   +
P1     11     12    PRD       SUB-ASSEM          2.0     X    *      I    *  I           - -    I    *    I    *    I   +
P1     15     18    PRD       SUB-ASS[illegible]         [illegible]     X    *      I    *                        *    I    *    I   +

P1     [illegible]                           FUNCTION TEST      1.0                              *    I
P1     24     [illegible]  WAD       ACCEPTANCE TEST    2.0     A                I   *    I    *    .    -    I    *    I   +
P1      1      2    TEC       DESIGN             5.0     CCCCCCCCCCCCCCCCCCCCCCCCC*    I    *    I    *    I    *    I
P1      2      3    TEC       PLANNING           3.0     X    *      I    *  I   AAAAAAAAAAAAAAA........................+
L                                                        X    *      I    *  I   *    I    *    I    *    I    *    I
P1      2     13    TEC       DESIGN BASE       10.0     X    *      I    *  I   CCCCCCCCCCCCCCCCCCCCCCCCCCCCCCCCCCCCCCC+
                              FRAME                      X    *      I    *  I   *    I    *    I    *    I    *    I
P1      2     14    TEC       DESIGN CIRCUITS    3.0     X    *      I    *  I   AAAAAAAAAAAAAAA.................. I
P1      6      7    TEC       TOOL DESIGN        6.0     X    *      I    *  I   *    AAAAAAAAAAAAAAAAAAAAAAAAAAAAAA....+
L                                                        X    *      I    *  I   *    I    *    I    *    I    *    I
P1     13     16    TEC       DESIGN COVERS      1.0     X    *      I    *  I   *    I    *    I    *    I    *    I   .+
```

Figure A2.3 Time analysis bar chart in report code and earliest start sequence

I C L 1900 SERIES PERT 22/10/69 OUTPUT SHEET NUMBER 43

PROJECT DB TEST NETWORK - PST 2.3 RUN 1 TIME NOW 5DEC66 PAGE 1

TIME-LIMITED RESOURCE ANALYSIS

S/P	PREC EVENT	SUCC EVENT	U I	REPORT CODE	DESCRIPTION	DUR	EARLIEST START	SCHED START	SCHED FINISH	LATEST FINISH	REM FLOAT	RESOURCES R1	R2
P3	1	2		TEC	DESIGN	3.0	5DEC66	5DEC66T	22DEC66	21DEC66	-.1	4A 3.0	2B 3.0
P1	1	2		TEC	DESIGN	5.0	5DEC66	5DEC66T	9JAN67	9JAN67	.0	3A 5.0	2B 5.0
P3	1	22		TEC	SPECIFY DRIVE MOTOR	1.0	5DEC66	5DBC66T	12DEC66	27JAN67	6.4	1A 1.0	
P3	22	28		PUR	OBTAIN MOTORS	14.0	12DEC66	12DEC66	18MAR67	5MAY67	6.4	*H 13.0 80H 1.0	
P3	1	18		QAD	DESGN & PLN RNNG-IN RIG	2.0	5DEC66	12DEC66	22DEC66	10MAR67	10.4	1A 2.0	1F 2.0
P3	2	4		TEC	PLAN CASTINGS	2.0	22DEC66	22DEC66	9JAN67	6JAN67	-.1	1A 2.0	1B 2.0
P3	2	5		TEC	PLAN COMPONENTS	3.0	22DEC66	22DEC66	16JAN67	17FEB67	4.4	1A 3.0	3B 3.0
P3	22	23		TEC	DESIGN CONTROLS	2.0	12DEC66	22DEC66	9JAN67	17FEB67	5.4	2A 2.0	1B 2.0
P3	18	19	A	PUR	OBTAIN B.O.PARTS	8.0	17DEC66	22DEC66	20FEB67	5MAY67	10.4	*H 7.0 160H 1.0	
P3	2	20		TEC	DESGN ASSEMBLY JIGS	1.0	22DEC66	22DEC66	2JAN67	23MAR67	12.0	1A 1.0	1B 1.0
[illegible]	20	16		PRD	MAKE JIGS	5.4	2JAN67	2JAN67	9JAN[illegible]		[illegible].1	1B 5.4	1F 5
P3	4	[illegible]		[illegible]	DESIGN TOOLS	4.0	9JAN[illegible]		6FEB67	17FEB67	1.4	[illegible]	1B 4.0
P1	2	4			LEAD	1.0	9JAN67	9JAN67	16JAN67	10FEB67	3.4		
P1	2	14		TEC	DESIGN CIRCUITS	3.0	9JAN67	9JAN67	30JAN67	24FEB67	3.4	2A 3.0	1B 3.0
P3	23	24		TEC	PLAN	2.0	22DEC66	9JAN67	23JAN67	3MAR67	5.4	1A 2.0	2B 2.0
P1	6	8			LEAD	2.0	16JAN67	16JAN67	30JAN67	3FEB67	.4		

Figure A2.4 Activity schedule for a time-limited project

I C L 1900 SERIES PERT 22/10/69 OUTPUT SHEET NUMBER 38

PROJECT DB TEST NETWORK - PST 2.3 RUN 1 TIME NOW 5DEC66 PAGE 1

RESOURCE-LIMITED ANALYSIS

S/P	PREC EVENT	SUCC EVENT	U I	REPORT CODE	DESCRIPTION	DUR	EARLIEST START	SCHED START	SCHED FINISH	LATEST FINISH	REM FLOAT	RESOURCES R1	R2
P3	1	2		TEC	DESIGN	3.0	5DEC66	5DEC66T	22DEC66	21DEC66	-.1	4A 3.0	2B 3.0
P1	1	2		TEC	DESIGN	5.0	5DEC66	5DEC66T	9JAN67	9JAN67	.0	3A 5.0	2B 5.0
P3	1	22		TEC	SPECIFY DRIVE MOTOR	1.0	5DEC66	5DEC66T	12DEC66	27JAN67	6.4	1A 1.0	
P3	22	28		PUR	OBTAIN MOTORS	14.0	12DEC66	12DEC66	18MAR67	5MAY67	6.4	*H 13.0 80H 1.0	
P3	20	16		PRD	MAKE JIGS	5.4	2JAN67	2JAN67 6FEB67	9JAN[illegible] 10MAR67	[illegible]MAY67	8.0	1B 5.4 60C 3.0 40C 2.0 *C .4	1F 5.4
P1	2	13		TEC	DESIGN BASE FRAME	10.0	9JAN67	9JAN67 23JAN67	16JAN67 23MAR67	18MAR67	-1.0	2A 10.0	1B 10.0
P3	4	14		PUR	OBTAIN CASTINGS	12.0	9JAN67	9JAN67	3APR67	31MAR67	-.1		
P1	2	6			LEAD	1.0	9JAN67	9JAN67	16JAN67	20JAN67	.4		
P3	4	13		TEC	DESIGN TOOLS	4.0	9JAN67	9JAN67	6FEB67	17FEB67	1.4	2A 4.0	1B 4.0
P1	2	4			LEAD	1.0	9JAN67	9JAN67	16JAN67	10FEB67	3.4		
P1	2	14		TEC	DESIGN CIRCUITS	3.0	9JAN67	9JAN67	30JAN67	24FEB67	3.4	2A 3.0	1B 3.0
P3	23	24		TEC	PLAN	2.0	22DEC66	9JAN67	23JAN67	3MAR67	5.4	1A 2.0	2B 2.0
P1 L	2	3		TEC	PLANNING	3.0	9JAN67	16JAN67	6FEB67	10MAR67	-1.0	1A 3.0	4B 3.0
P1	6	8			LEAD	2.0	16JAN67	16JAN67	30JAN67	3FEB67	.4		

Figure A2.5 Activity schedule for a resource-limited project

C.P.P.—7

```
I C L  1900 SERIES PERT                           10/12/69                             OUTPUT SHEET NUMBER   31
                  PROJECT  DB   TEST NETWORK - PST 2.3                      RUN   1  TIME-NOW  5DEC66  PAGE   5
                                TIME-LIMITED RESOURCE ANALYSIS
             B  /PLANNERS                0         10        20        30        40        50        60
DATE              AV    REQ    REM       I.........I.........I.........I.........I.........I.........I     DATE

  *                                                                                                            *
 5DEC66            8     4      4        X####////XXXXXXXXXXXXXXXXXXXXXXXXXXXXXXXXXXXXXXXXXXXXXXXXXXXX      .0
 6DEC66            8     4      4        I   #   / I         I         I         I         I         I      .1
 7DEC66            8     4      4        I   #   / I         I         I         I         I         I      .2
 8DEC66            8     4      4        I   #   / I         I         I         I         I         I      .3
 9DEC66            8     4      4        I   #   / I         I         I         I         I         I      .4

12DEC66            8     4      4        I   #   / I         I         I         I         I         I     1.0
13DEC66            8     4      4        I   #   / I         I         I         I         I         I     1.1
14DEC66            8     4      4        I   #   / I         I         I         I         I         I     1.2
15DEC66            8     4      4        I   #   / I         I         I         I         I         I     1.3
16DEC66            8     4      4        I   #   / I         I         I         I         I         I     1.4
17DEC66            8     4      4        I   #   / I         I         I         I         I         I     1.5
18DEC66            8     4      4        I   #   / I         I         I         I         I         I     1.6
19DEC66            8     4      4        I   #   / I         I         I         I         I         I     2.0
20DEC66            8     4      4        I   #   / I         I         I         I         I         I     2.1
21DEC66            8     4      4        I   #   / I         I         I         I         I         I     2.2
22DEC66            8     8      0        I    #### I         I         I         I         I         I     2.3
23DEC66            8     8      0        I       # I         I         I         I         I         I     2.4
28DEC66            8     8      0        I       # I         I         I         I         I         I     3.2
29DEC66            8     8      0        I       # I         I         I         I         I         I     3.3
30DEC66            8     8      0        I       # I         I         I         I         I         I     3.4
 2JAN67            8     8      0        I       # I         I         I         I         I         I     4.0
 3JAN67            8     8      -                # I                             I         I         .     4.1
  ''AN67           8     P                                                                 I               /
                                         I       #                     I         I                   I
26JAN67            8     8      0        I       # I         I         I         I         I         I     7..
27JAN67            8     8      0        I       # I         I         I         I         I         I     7.4
30JAN67            8     7      1        I      #/ I         I         I         I         I         I     8.0
31JAN67            8     7      1        I      #/ I         I         I         I         I         I     8.1
 1FEB67            8     7      1        I      #/ I         I         I         I         I         I     8.2
 2FEB67            8     7      1        I      #/ I         I         I         I         I         I     8.3
 3FEB67            8     8      0        I       # I         I         I         I         I         I     8.4
 6FEB67            8     7      1        I      #/ I         I         I         I         I         I     9.0
```

Figure A2.6 Resource histogram for a time-limited run

```
I C L  1900 SERIES PERT                          22/10/69                                  OUTPUT SHEET NUMBER   48

                    PROJECT  DB   TEST NETWORK : PST 2 3                        RUN   1  TIME:NOW  5DEC66  PAGE   1

                                  RESOURCE-LIMITED ANALYSIS

        A   /DESIGNERS                0         10        20        30        40        50        60
DATE           AV     REQ     REM     I.........I.........I.........I.........I.........I.........I              DATE

  *                                                                                                                  *
 5DEC66         8       8       0     X########XXXXXXXXXXXXXXXXXXXXXXXXXXXXXXXXXXXXXXXXXXXXXXXXXXXX              .0
 6DEC66         8       8       0     I       # I         I         I         I         I         I              .1
 7DEC66         8       8       0     I       # I         I         I         I         I         I              .2
 8DEC66         8       8       0     I       # I         I         I         I         I         I              .3
 9DEC66         8       8       0     I       # I         I         I         I         I         I              .4
12DEC66         8       8       0     I       # I         I         I         I         I         I             1.0
13DEC66         8       8       0     I       # I         I         I         I         I         I             1.1
14DEC66         8       8       0     I       # I         I         I         I         I         I             1.2
15DEC66         8       8       0     I       # I         I         I         I         I         I             1.3
16DEC66         8       8       0     I       # I         I         I         I         I         I             1.4
17DEC66         8       8       0     I       # I         I         I         I         I         I             1.5
18DEC66         8       8       0     I       # I         I         I         I         I         I             1.6
19DEC66         8       8       0     I       # I         I         I         I         I         I             2.0
20DEC66         8       8       0     I       # I         I         I         I         I         I             2.1
21DEC66         8       8       0     I       # I         I         I         I         I         I             2.2
22DEC66         8       8       0     I       # I         I         I         I         I         I             2.3
23DEC66         8       8       0             # I                   I         I         I         I             2.4
28DE[illegible] 8       8                                                     I         I
                8

                        6       2           # / I         I         I                   .         I             6.2
19JAN67         3       6       2     I     # / I         I         I         I         I         I             6.3
20JAN67         8       6       2     I     # / I         I         I         I         I         I             6.4
23JAN67         8       7       1     I      #/ I         I         I         I         I         I             7.0
24JAN67         8       7       1     I      #/ I         I         I         I         I         I             7.1
25JAN67         8       7       1     I      #/ I         I         I         I         I         I             7.2
26JAN67         8       7       1     I      #/ I         I         I         I         I         I             7.3
27JAN67         8       7       1     I      #/ I         I         I         I         I         I             7.4
30JAN67         8       8       0     I       # I         I         I         I         I         I             8.0
31JAN67         8       8       0     I       # I         I         I         I         I         I             8.1
 1FEB67         8       8       0     I       # I         I         I         I         I         I             8.2
 2FEB67         8       8       0     I       # I         I         I         I         I         I             8.3
 3FEB67         8       8       0     I       # I         I         I         I         I         I             8.4
 6FEB67         8       5       3     I    ###/ I         I         I         I         I         I             9.0
```

Figure A2.7 Resource histogram for a resource-limited project

```
I C L  1900 SERIES PERT                          21/11/69                            OUTPUT SHEET NUMBER   24
                    PROJECT  DB   TEST NETWORK - PST 2.3                  RUN    0  TIME-NOW  5DEC66  PAGE    1
                                  TIME-LIMITED ANALYSIS WITH OVERLOAD
              A   /DESIGNERS                0         10        20        30        40        50        60
DATE               AV     REQ     REM       I.........I.........I.........I.........I.........I.........I            DATE

  *                                                                                                                     *
 5DEC66            8       8       0        X#######XXXXXXXXXXXXXXXXXXXXXXXXXXXXXXXXXXXXXXXXXXXXXXXXXXXXX            .0
 6DEC66            8       8       0        I######## I         I         I         I         I         I            .1
 7DEC66            8       8       0        I######## I         I         I         I         I         I            .2
 8DEC66            8       8       0        I######## I         I         I         I         I         I            .3
 9DEC66            8       8       0        I######## I         I         I         I         I         I            .4
12DEC66            8       8       0        I######## I         I         I         I         I         I           1.0
12DEC66            8       8       0        I######## I         I         I         I         I         I           1.1
14DEC66            8       8       0        I######## I         I         I         I         I         I           1.2
15DEC66            8       8       0        I######## I         I         I         I         I         I           1.3
16DEC66            8       8       0        I######## I         I         I         I         I         I           1.4
17DEC66            8       8       0        I######## I         I         I         I         I         I           1.5
18DEC66            8       8       0        I######## I         I         I         I         I         I           1.6
19DEC66            8       8       0        I######## I         I         I         I         I         I           2.0
20DEC66            8       8       0        I######## I         I         I         I         I         I           2.1
21DEC66            8       8       0        I######## I         I         I         I         I         I           2.2
22DEC66            8       8       0        I######## I         I         I         I         I         I           2.3
23DEC66            8       8       0        I######## I         I         I         I         I         I           2.4
28DEC66            8       8       0        I######## I         I         I         I         I         I           3.2
?9DEC66            8       8       0        I######## I         I         I         I         .         I           3.3
   -C66            8       8       0                  -         I         I                                          3

                                   3        I##..               .                             I         I
23JAN67            8      10      -2        I######..           I         I         I         I         I           7.0
24JAN67            8      10      -2        I########--         I         I         I         I         I           7.1
25JAN67            8      10      -2        I########--         I         I         I         I         I           7.2
26JAN67            8      10      -2        I########--         I         I         I         I         I           7.3
27JAN67            8      10      -2        I########--         I         I         I         I         I           7.4
30JAN67            8       7       1        I#######/ I         I         I         I         I         I           8.0
31JAN67            8       7       1        I#######/ I         I         I         I         I         I           8.1
 1FEB67            8       7       1        I#######/ I         I         I         I         I         I           8.2
 2FEB67            8       7       1        I#######/ I         I         I         I         I         I           8.3
 3FEB67            8       7       1        I#######/ I         I         I         I         I         I           8.4
 6FEB67            8       5       3        I#####/// I         I         I         I         I         I           9.0
```

Figure A2.7a Resource histogram for a time-limited run : alternative format

I C L 1900 SERIES PERT 22/10/69 OUTPUT SHEET NUMBER 1041

PROJECT DB TEST NETWORK - PST 2.3 RUN 1 TIME-NOW 5DEC66 PAGE 1

RESOURCE TABLE

	A /DESIGNERS			B /PLANNERS			C /TOOL ROOM HOURS			D /MC SHOP HOURS			
DATE	AV	REQ	REM	AV	REQ	REM	AV	REQ	REM	AV	REQ	REM	DATE
5DEC66	8	8	0	8	4	4	200	0	200	400	0	400	.0
22DEC66	8	8	0	8	8	0	200	0	200	400	0	400	2.3
2JAN67	8	7	1	8	8	0	200	60	140	400	0	400	4.0
9JAN67	8	8	0	8	8	0	200	0	200	400	0	400	5.0
16JAN67	8	6	2	8	8	0	200	0	200	400	0	400	6.0
23JAN67	8	7	1	8	8	0	200	10	190	400	40	360	7.0
30JAN67	8	8	0	8	8	0	200	10	190	400	60	340	8.0
6FEB67	8	5	3	8	7	1	200	130	70	400	120	280	9.0
13FEB67	8	5	3	8	7	1	200	190	10	400	80	320	10.0
[illegible]FEB67	8	5	3	8	7	1	200	190	10	400	100	300	11.0
[illegible]	8	5	3	8			[illegible]	130	70	[illegible]	140	260	1[illegible]
10APR[illegible]					3	5			[illegible]	400			[illegible]
17APR67	8	0	8	8	2	6	200	190	10	400	120	280	19.0
24APR67	8	0	8	8	3	5	200	190	10	400	120	280	20.0
1MAY67	8	0	8	8	2	6	200	140	60	400	260	140	21.0
8MAY67	8	0	8	8	2	6	200	120	80	400	260	140	22.0
9MAY67	8	0	8	8	2	6	200	120	80	400	380	20	22.1

Figure A2.8 Resource table for normal resources

```
.I C L  1900 SERIES PERT                         26/11/69                                    OUTPUT SHEET NUMBER   25

                    PROJECT  DB   TEST NETWORK - PST 2.3                          RUN   1  TIME NOW  5DEC66  PAGE   1

 BAR CHART OF TIME-LIMITED RESOURCE ANALYSIS

                                                           5          19       2        16        30        13        27
S/P  PREC    SUCC U  REPORT    DESCRIPTION         DUR  DEC66       DEC66    JAN67    JAN67     JAN67     FEB67     FEB67
     EVENT  EVENT I   CODE                               X....*......I....*..I....*....I....*....I....*....I....*....I...
P3     1      2   TEC       DESIGN              3.0  -CCCCCCCCCCCCCCC  *  I    *    I    *    I    *    I    *    I
P1     1      2   TEC       DESIGN              5.0  CCCCCCCCCCCCCCCCCCCCCCCCC*    I    *    I    *    I    *    I
P3     1     22   TEC       SPECIFY DRIVE       1.0  AAAAA...............................  I    *    I    *    I
                            MOTOR                    X    *      I    *  I    *    I    *    I    *    I    *    I
P3    22     28   PUR       OBTAIN MOTORS      14.0  X    AAAAAAAAAAAAAAAAAAAAAAAAAAAAAAAAAAAAAAAAAAAAAAAAAAAAAAA+
P3     1     18   QAD       DESGN & PLN         2.0  X....AAAAAAAAAA........................................+
                            RNNG-IN RIG              X    *      I    *  I    *    I    *    I    *    I    *    I

P3    20                                             X    *
P3     4     14   PUR       OBTAIN CASTINGS    12.0  X    *      I
P1     2      3   TEC       PLANNING            3.0  X    *      I    *  I    AAAAAAAAAAAAAAA....................
L                                                    X    *      I    *  I    *    I    *    I    *    I    *    I
P1     2     13   TEC       DESIGN BASE        10.0  X    *      I    *  I    CCCCCCCCCCCCCCCCCCCCCCCCCCCCCCCCCCCCCCC+
                            FRAME                    X    *      I    *  I    *    I    *    I    *    I    *    I
P1     2      6             LEAD                1.0  X    *      I    *  I    AAAAA.... *    I    *    I    *    I
P3     4     13   TEC       DESIGN TOOLS        4.0  X    *      I    *  I    AAAAAAAAAAAAAAAAAAAA......... *    I
P1     2      4             LEAD                1.0  X    *      I    *  I    AAAAA.................... I    *    I
P1     2     14   TEC       DESIGN CIRCUITS     3.0  X    *      I    *  I    AAAAAAAAAAAAAAA................... I
P3    23     24   TEC       PLAN                2.0  X    *      I..........AAAAAAAAAA..............................
P1     6      8             LEAD                2.0  X    *      I    *  I    *    AAAAAAAAAA.... *    I    *    I
P1     4      5   PUR       OBTAIN RAW MAT'S    9.0              I    *  I    *    AAAAAAAAAAAAAAAAAAAAAAAAAAAAAAAAAAA+
                                                                      *                   I    *    I

                 19 B PRD       MAKE PA..           6.0  X    *              .....................
       3      7             LAG                 2.0  X    *      I    *  I    *    I    *    AAAAAAAAAA.............
.1     3      5             LAG                 4.0  X    *      I    *  I    *    I    *    AAAAAAAAAAAAAAAAAAAA....+
P3    24     25 A PRD       MAKE PARTS          5.0  X    *      I    *  I    ..............AAAAAAAAAAAAAAAAAAAAAAAAA+
P1    14     15 B PRD       MAKE SPECIAL        5.0  X    *      I    *  I    *    I    *    AAAAAAAAAAAAAAAAAAAAAAAAA+
                            ITEMS                    X    *      I    *  I    *    I    *    I    *    I    *    I
P1     8      9   PRD       TOOL MANUFACTURE   10.0  X    *      I    *  I    *    I    *    ....CCCCCCCCCCCCCCCCCCCC+
L                                                    X    *      I    *  I    *    I    *    I    *    I    *    I
P3    13     14   PRD       MAKE TOOLS          6.0  X    *      I    *  I    *    I    *    I    AAAAAAAAAAAAAAAAAAA+
P1    10     11   PRD       P/P MANU            8.0  X    *      I    *  I    *    I    *    I    *    I    ....A.AAA+
L                                                    X    *      I    *  I    *    I    *    I    *    I    *    I
```

Figure A2.9 Bar chart of time-limited resource analysis

```
I C L  1900 SERIES PERT                                26/11/69                                     OUTPUT SHEET NUMBER   27
                         PROJECT  DB   TEST NETWORK - PST 2.3                          RUN    1  TIME NOW  5DEC66  PAGE    1
BAR CHART OF TIME-LIMITED RESOURCE ANALYSIS
      13         28       10        24         8        22        5        19         3        17        31        14
    MAR67      MAR67    APR67    'APR67     MAY67     MAY67    JUN67     JUN67     JUL67     JUL67     JUL67     AUG67
 .*....I......*...I...*....I....*....I....*....I....*....I....*...I....*.....I....*....I....*....I....*....I...*....I
- *    I      *   I   *    I    *    I    *    I    *    I    *   I    *     I    *    I    *    I    *    I   *    I
- *    I      *   I   *    I    *    I    *    I    *    I    *   I    *     I    *    I    *    I    *    I   *    I
- *    I      *   I   *    I    *    I    *    I    *    I    *   I    *     I    *    I    *    I    *    I   *    I
  *    I      *   I   *    I    *    I    *    I    *    I    *   I    *     I    *    I    *    I    *    I   *    I
-AAAAAAAAAAA.................................. I    *    I    *   I    *     I    *    I    *    I    *    I   *    I
-....  I      *   I   *    I    *    I    *    I    *    I    *   I    *     I    *    I    *    I    *    I   *    I
  *    I      *   I   *    I    *    I    *    I    *    I    *   I    *     I    *    I    *    I    *    I   *    I
- *    I      *   I   *    I    *    I    *    I    *    I    *   I    *     I    *    I    *    I    *    I   *    I
- *    I      *   I   *    I    *    I    *    I    *    I    *   I    *     I    *    I    *    I    *    I   *    I
- *    I      *   I   *    I    *    I    *    I    *    I    *   I    *     I    *    I    *    I    *    I   *    I
-............................................. I    *    I    *   I    *     I    *    I    *    I    *    I   *    I
-................ I   *    I    *    I    *    I    *    I    *   I    *     I    *    I    *    I    *    I   *    I
  *    I      *   I   *    I    *    I    *    I    *    I    *   I    *     I    *    I    *    I    *    I   *    I
-.................A..AAAAAAAAA................ I    *    I    *   I    *     I    *    I    *    I    *    I   *    I
-CCCCCCCCCCCCCCCCCCCCC*    I    *    I    *    I    *    I    *   I    *     I    *    I    *    I    *    I   *    I
-.....  I      *   I   *    I    *    I    *    I    *    I    *   I    *     I    *    I    *    I    *    I   *    I
  *    I      *   I   *    I    *    I    *    I    *    I    *   I    *     I    *    I    *    I    *    I   *    I
-CCCCCCCCCCC  *   I   *    I    *    I    *    I    *    I    *   I    *     I    *    I    *    I    *    I   *    I
  *    I      *   I   *    I    *    I    *    I    *    I    *   I    *     I    *    I    *    I    *    I   *    I
- *    I      *   I   *    I    *    I    *    I    *    I    *   I    *     I    *    I    *    I    *    I   *    I
- *    I      *   I   *    I    *    I    *    I    *    I    *   I    *     I    *    I    *    I    *    I   *    I
- *    I      *   '   *    I    *    I    *    I    *    I    *   I    *     I    *    I    *    I    *    I   *    I
- *    I              *    I    *    I    *    I    *    I    *   I    *     I    *    I    *    I    *    I   *    I
                      *    I    *    .         .    *    I    *   I    *     I    *    I    *    I    *    I   *    I
                           I                                  -   I    *     I    -              .    *    I   *    I

-....                       --- -    I    *    I    *                             .    I    *    -              -    I
-A............    ............................ I    *    I    -              I    *    I    *    I    *    I   *    I
-A................................................. *    I    *   I    *     I    *    I    *    I    *    I   *    I
  *    I      *   I   *    I    *    I    *    I    *    I    *   I    *     I    *    I    *    I    *    I   *    I
-CCCCCCCCCCCCCCCCCCCCCCCCCCCCCC *    I    *    I    *    I    *   I    *     I    *    I    *    I    *    I   *    I
  *    I      *   I   *    I    *    I    *    I    *    I    *   I    *     I    *    I    *    I    *    I   *    I
-A.....AAAAAAAAAA....  *    I    *    I    *    I    *    I    *   I    *     I    *    I    *    I    *    I   *    I
-AAA.AAAAAA.AAAAAA.AAAAAA.CCCCCCCCCCCCCC  *    I    *    I    *   I.   *     I    *    I    *    I    *    I   *    I
  *    I      *   I   *    I    *    I    *    I    *    I    *   I    *     I    *    I    *    I    *    I   *    I
```

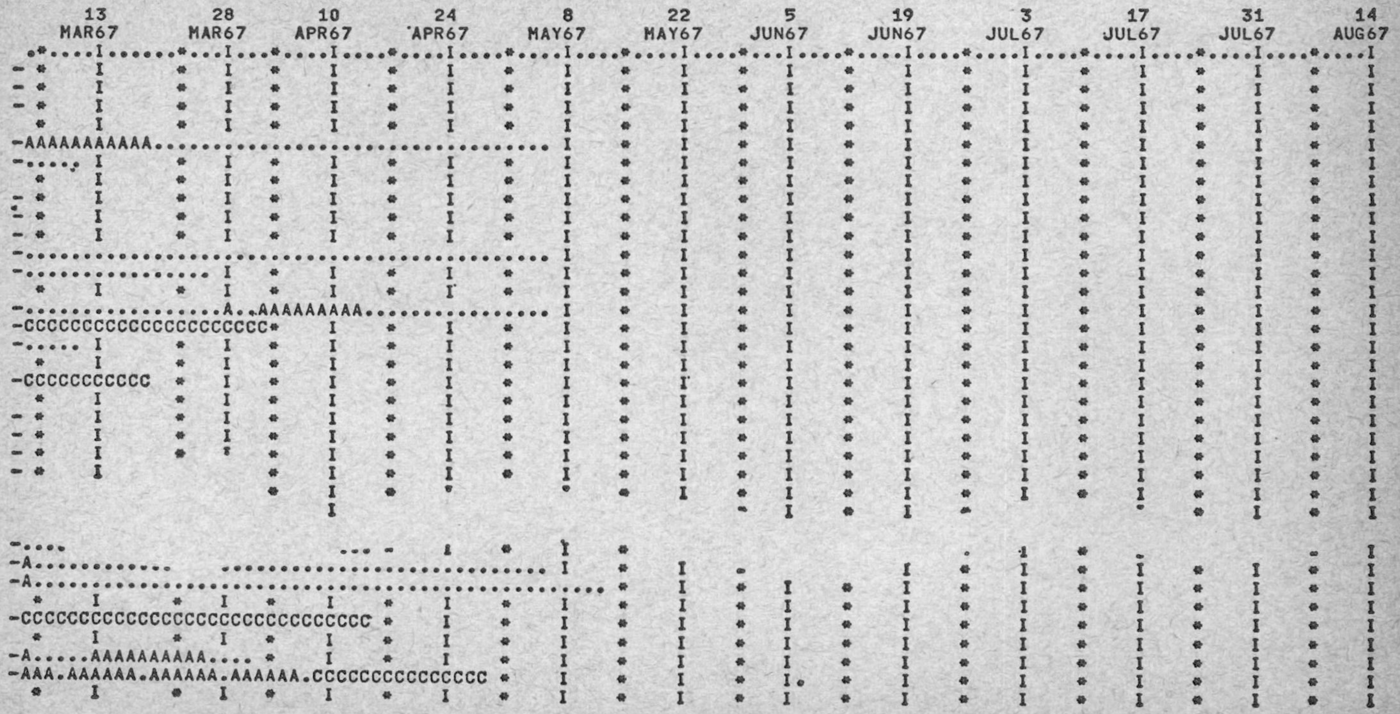

Figure A2.9a Bar chart of time-limited resource analysis (continued)

```
I C L  1900 SERIES PERT                              14/12/69                            OUTPUT SHEET NUMBER   23

                  PROJECT  DB   TEST NETWORK - PST 2.3                          RUN   1  TIME-NOW  5DEC66  PAGE   2

                                RESOURCE-LIMITED COST

        TOT /TOTAL PROJECT                0      10000     20000     30000     40000     50000     60000
DATE    PLAN COST CUM PLAN                I.........I.........I.........I.........I.........I.........I           DATE

 7FEB67       742     9574                I        PI         I         I         I         I         I            9.1
 8FEB67       742    10316                I         P         I         I         I         I         I            9.2
 9FEB67       742    11058                I         IP        I         I         I         I         I            9.3
10FEB67       742    11800                I         IP        I         I         I         I         I            9.4
13FEB67      1002    12802                I         I P       I         I         I         I         I           10.0
14FEB67      1002    13804                I         I  P      I         I         I         I         I           10.1
15FEB67      1002    14806                I         I   P     I         I         I         I         I           10.2
16FEB67      1002    15808                I         I    P    I         I         I         I         I           10.3
17FEB67      1002    16810                I         I     P   I         I         I         I         I           10.4
20FEB67       894    17704                I         I      P  I         I         I         I         I           11.0
21FEB67       894    18598                I         I       P I         I         I         I         I           11.1
22FEB67       894    19492                I         I        PI         I         I         I         I           11.2
23FEB67       894    20386                I         I         P         I         I         I         I           11.3
24FEB67       894    21280                I         I         IP        I         I         I         I           11.4
27FEB67       800    [illegible]          I         I         [illegible]         I         I         I         I           12
28FEB[illegible]

          [illegible]4    36188                                                                                   14.5
19MAR67      1590    37778                I         I         I         I       P I         I         I           14.6
20MAR67      1584    39362                I         I         I         I        PPI        I         I           15.0
21MAR67      1590    40952                I         I         I         I         P         I         I           15.1
22MAR67      1596    42548                I         I         I         I         IPP       I         I           15.2
23MAR67      1290    43838                I         I         I         I         I  P      I         I           15.3
28MAR67       774    44612                I         I         I         I         I   P     I         I           16.1
29MAR67      1302    45914                I         I         I         I         I    P    I         I           16.2
30MAR67      1302    47216                I         I         I         I         I     PP  I         I           16.3
31MAR67      1302    48518                I         I         I         I         I       P I         I           16.4
 3APR67       914    49432                I         I         I         I         I        PI         I           17.0
 4APR67       914    50346                I         I         I         I         I         P         I           17.1
 5APR67       914    51260                I         I         I         I         I         IP        I           17.2
 6APR67       914    52174                I         I         I         I         I         I P       I           17.3
 7APR67       914    53088                I         I         I         I         I         I  P      I           17.4
10APR67      1056    54144                I         I         I         I         I         I   P     I           18.0
11APR67      1056    55200                I         I         I         I         I         I    P    I           18.1
12APR67      1056    56256                I         I         I         I         I         I     P   I           18.2
```

Figure A2.10 Planned cost output (suitable for simulating the effect of choosing a particular schedule)

P = PLAN COST = the cost to be incurred for the resource during the time period

CUM PLAN = cumulative plan cost; the accumulated planned cost for the resource to date

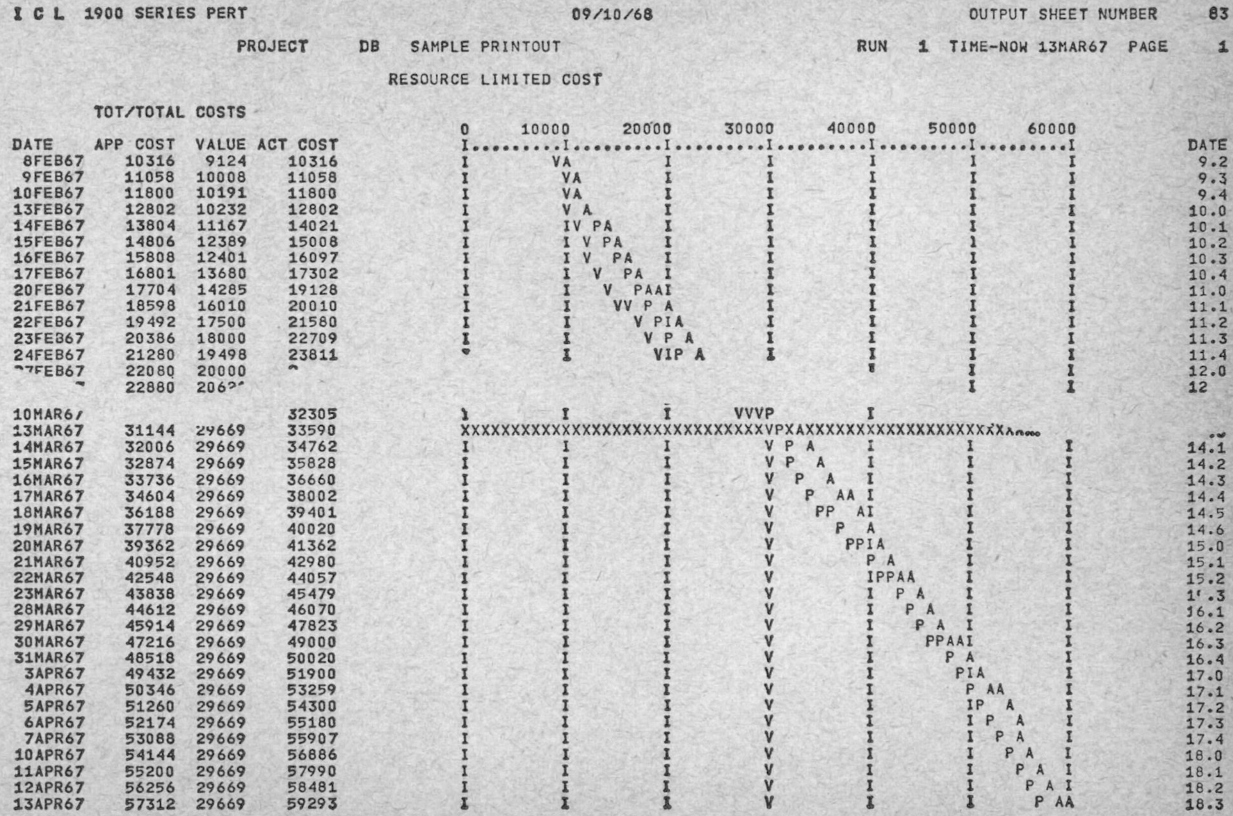

I C L 1900 SERIES PERT 09/10/68 OUTPUT SHEET NUMBER 83

PROJECT DB SAMPLE PRINTOUT RUN 1 TIME-NOW 13MAR67 PAGE 1

RESOURCE LIMITED COST

TOT/TOTAL COSTS

DATE	APP COST	VALUE	ACT COST	DATE
8FEB67	10316	9124	10316	9.2
9FEB67	11058	10008	11058	9.3
10FEB67	11800	10191	11800	9.4
13FEB67	12802	10232	12802	10.0
14FEB67	13804	11167	14021	10.1
15FEB67	14806	12389	15008	10.2
16FEB67	15808	12401	16097	10.3
17FEB67	16801	13680	17302	10.4
20FEB67	17704	14285	19128	11.0
21FEB67	18598	16010	20010	11.1
22FEB67	19492	17500	21580	11.2
23FEB67	20386	18000	22709	11.3
24FEB67	21280	19498	23811	11.4
~7FEB67	22080	20000		12.0
	22880	2062'		12
10MAR6/			32305	
13MAR67	31144	29669	33590	
14MAR67	32006	29669	34762	14.1
15MAR67	32874	29669	35828	14.2
16MAR67	33736	29669	36660	14.3
17MAR67	34604	29669	38002	14.4
18MAR67	36188	29669	39401	14.5
19MAR67	37778	29669	40020	14.6
20MAR67	39362	29669	41362	15.0
21MAR67	40952	29669	42980	15.1
22MAR67	42548	29669	44057	15.2
23MAR67	43838	29669	45479	1'.3
28MAR67	44612	29669	46070	16.1
29MAR67	45914	29669	47823	16.2
30MAR67	47216	29669	49000	16.3
31MAR67	48518	29669	50020	16.4
3APR67	49432	29669	51900	17.0
4APR67	50346	29669	53259	17.1
5APR67	51260	29669	54300	17.2
6APR67	52174	29669	55180	17.3
7APR67	53088	29669	55907	17.4
10APR67	54144	29669	56886	18.0
11APR67	55200	29669	57990	18.1
12APR67	56256	29669	58481	18.2
13APR67	57312	29669	59293	18.3

Figure A2.11 General cost report (user compare actual cost so far, and outlook on the basis of current achievements with planned costs of original schedule). APP COST = planned cost to date according to approved schedule. V = VALUE = value of work to date; the original approved work actually completed. A = ACT COST = actual cost for dates up to TIME NOW, and for later dates adjusted or revised costs. P = PLANNED COST = cost incurred for the resource during the time period.

Appendix Three List of firms offering facilities for network analysis

[provided by, and printed by kind permission of, Computer Services and Bureaux Association (COSBA), Leicester House, 8 Leicester Street, London, WC2H 7BN, telephone number 01-437 0678 (whose secretary will be pleased to provide any further information)]

Baric Computing Services Limited
Kidsgrove, Stoke-on-Trent, ST7 ITL
Tel: 0782 29681

Cambridge Computer Services Limited
Jupiter House, Station Road, Cambridge, CB1 2JY
Tel: 0223 66111

Capital Cities Computer Centres Limited
46 Clarendon Road, Watford, Hertfordshire
Tel: 92 38321

Computeraid Limited
Ladymead, Guildford, Surrey
Tel: 0483 64691

Computer Projects Limited
8–16 Great New Street, London, E.C.4
Tel: 01-585 9205

Computer Time Brokers Limited
38 Park Street, London, W1Y 3PF
Tel: 01-499 5948

Data Sciences International Limited
DSI House, Sheepscar Street South, Leeds, Yorkshire
Tel: 0532 41541

Dychurch Business Services Limited
District Bank Chambers, 12 Gold Street, Northampton
Tel: 0604 33394

Feni Data Services Limited
17 Hinton Road, Bournemouth, Hampshire BH1 2EE
Tel: 0202 28455

Hoskyns Group Limited
Boundary House, Furnival Street, London, E.C.4
Tel: 01-242 1951

IBM United Kingdom Limited
389 Chiswick High Road, London, W.4
Tel: 01-995 1441

Inbucon Computer Bureaux Limited
Hadley House, 79 Uxbridge Road, Ealing, London, W.5
Tel: 01-579 1661

ISIS Computer Services Limited
Forum House, 15–18 Lime Street, London, E.C.3
Tel: 01-623 0111

ITT Data Services
153–155 East Barnet Road, East Barnet, Hertfordshire
Tel: 01-440 5161

Kent Data Services Limited
134–154 Biscot Road, Luton, Bedfordshire
Tel: Luton 32483

Laing Computing Services
Page Street, London, N.W.7
Tel: 01-906 5065

Leasco Response Limited
197 Knightsbridge, London, S.W.7
Tel: 01-584 7040

London University Computing Services Limited
39 Gordon Square, London, W.1.
Tel: 01-387 3421

Lowndes-Ajax Computer Service Limited
Carolyn House, Dingwall Road, Croydon, Surrey, CR9 2XG
Tel: 01-686 3661

Management Computing Services Limited
Adelphi, 1 John Adam Street, London, W.C.2
Tel: 01-930 5833

NCB Computer Power
Bridgtown, Cannock, Staffordshire
Tel: Cannock 2581

Northamptonshire Computer Bureau Limited
Station Road, Burton Latimer, Nr. Kettering, Northamptonshire
Tel: 0536-72 3651

Randax EDP Limited
10–14 Macklin Street, London, W.C.2
Tel: 01-242 6291

Sanaco Computer Services
Rockville Road, Birmingham 8
Tel: 021-327 3831

Scientific Control Systems Limited
Sanderson House, 49–57 Berners Street, London, W.1
Tel: 01-580 5599

SIA Limited
Ebury Gate, 23 Lower Belgrave Street, London, S.W.1
Tel: 01-730 4544

Time Sharing Limited
179–193 Great Portland Street, London, W.1
Tel: 01-637 1355

University Computing Company (Great Britain) Limited
UCC Computer Centre, 143 Bromsgrove Street, Birmingham 5
Tel: 021-692 1041

Appendix Four **Further reading**

Recommended reading

Battersby, A. *Network Analysis for Planning and Scheduling.* 3rd edition. London, Macmillan; New York, St. Martins, 1970. pp. 332.

Woodgate, H. S. *Planning by Network.* 2nd edition. London, Business Publications Ltd., 1967. pp. 406.

Other books on critical path planning

Armstrong-Wright, A. T. *Critical Path Method : introduction and practice.* Harlow, Loongmans, 1969. pp. 26.

Baboulene, Bernard. *Critical Path Made Easy.* London, Duckworth, 1970. pp. 101.

Barnetson, Paul. *Critical Path Planning : present and future techniques.* London, Newnes Butterworth, 1968. pp. 102.

Couldery, Frederick A. J. *Critical Path Analysis.* Hove, Editype Ltd., 1967. pp. 32.

Holden, I. R. and McIlroy, P. K. *Network Planning in Management Control Systems.* London, Hutchinson Educational, 1970. pp. 119.

Kaufmann, A., and Desbazeille, G. *The Critical Path Method: application of the PERT method and its variants to production and study programs.* New York and London, Gordon & Breach, 1969. pp. 187.

Lambourn, Simon. *Network Analysis in Project Management : a basic introduction to PERT and other critical path methods.* London, Industrial and Commercial Techniques Ltd., 1965. pp. 62.

Larkin, J. A. *Card Network Planning.* London, Scottish Academic Press, distributed by Chatto and Windus, 1970. pp. 31.

Lockyer, K. G. *An Introduction to Critical Path Analysis.* 3rd edition. London, Pitman, 1969. pp. 230.

Mulvaney, J. E. *Analysis Bar Charting: a simplified critical path analysis technique.* London, Iliffe, 1969. pp. 100.

Lowe, C. W. *Critical Path Analysis by Bar Chart: the new role of job progress charts.* 2nd edition. London, Business Books, 1969. pp. 210.

Mulvaney, J. E. *The Use of Network Analysis in Marketing.* London, Institute of Marketing, 1969. pp. 47.

Simms, Alfred G., and Britten, John R. *Project Network Analysis and Critical Path.* Brighton, Machinery Publishing, 1969. pp. 98.

Wiest, Jerome D., and Levy, Ferdinand K. *A Management Guide to PERT/CPM.* London Prentice Hall, 1969. pp. 175.

Answers to Exercises in drawing arrow diagrams

Exercise 1 (page 9)

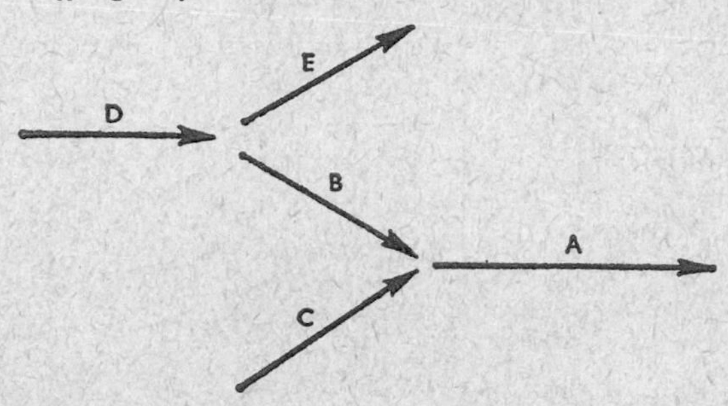

Exercise 2 (page 9)

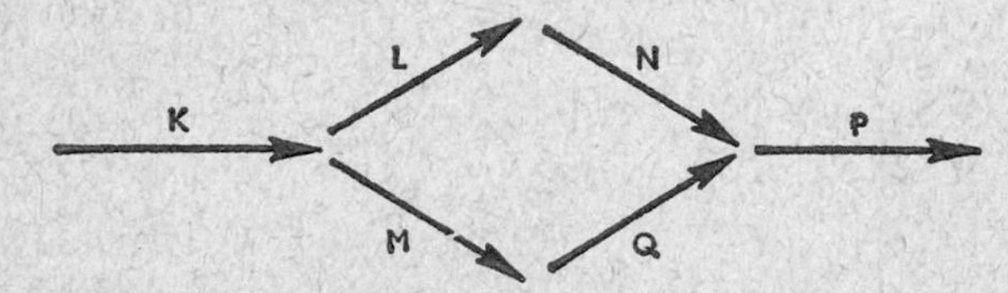

Exercise 3 (page 16)

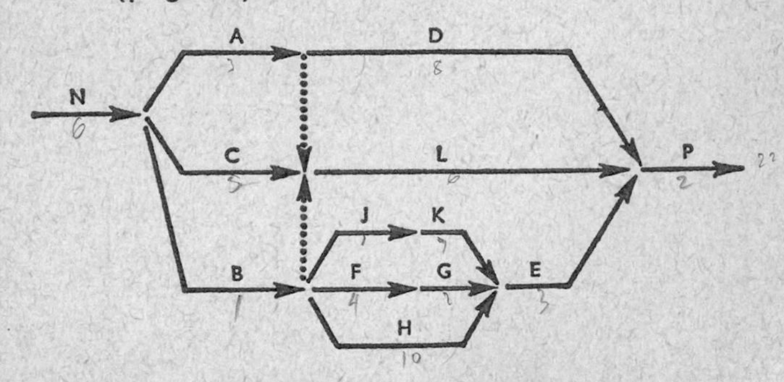

Exercise 4 (page 19)

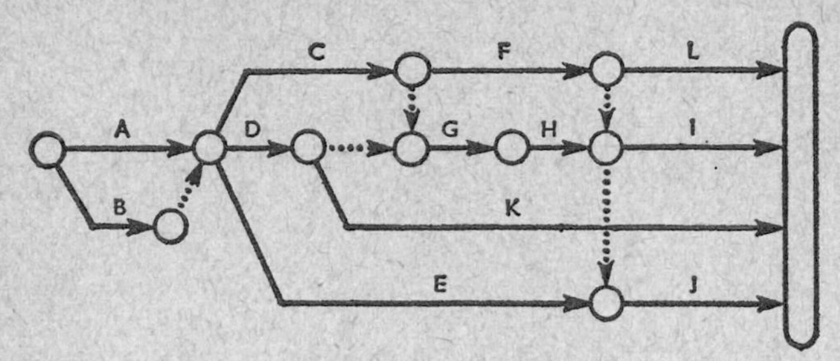

Exercise 5 (page 20)

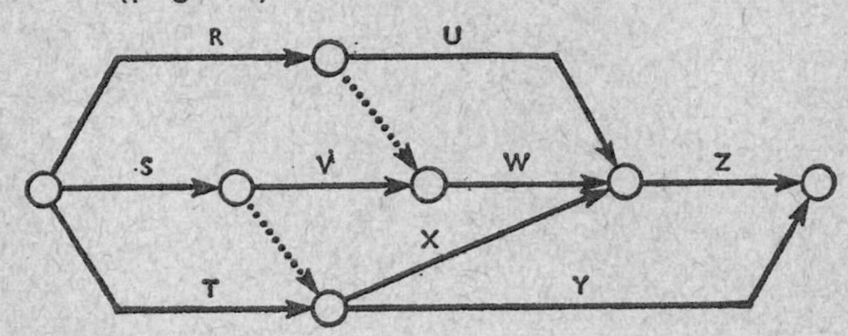

Exercise 6 (page 20)

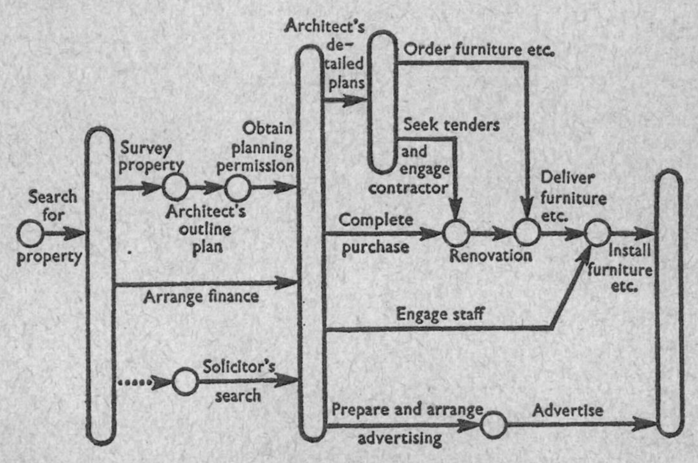

Exercise 7 (page 47)

Project duration = 22 days.

Critical events : 1, 2, 5, 8, 9, 10.

Critical activities : N, B, H, E, P.

Activity		Duration	EST	EFT	LST	LFT	TF	FF
N	1 – 2	6	0	6	0	6	0	0
A	2 – 3	3	6	9	9	12	3	0
C	2 – 4	5	6	11	9	14	3	0
B	2 – 5	1	6	7	6	7	0	0
D	3 – 9	8	9	17	12	20	3	3
L	4 – 9	6	11	17	14	20	3	3
J	5 – 6	1	7	8	9	10	2	0
F	5 – 7	4	7	11	11	15	4	0
H	5 – 8	10	7	17	7	17	0	0
K	6 – 8	7	8	15	10	17	2	2
G	7 – 8	2	11	13	15	17	4	4
E	8 – 9	3	17	20	17	20	0	0
P	9 – 10	2	20	22	20	22	0	0

Exercise 8 (page 47)

Activity		Duration	EST	EFT	LST	LFT	TF	FF	
B	1 – 2	4	0	4	0	4	0	0	*
A	1 – 3	3	0	3	1	4	1	1	
Dum.	2 – 3	0	4	4	4	4	0	0	*
C	3 – 4	2	4	6	8	10	4	0	
D	3 – 5	6	4	10	4	10	0	0	*
E	3 – 10	2	4	6	18	20	14	12	
Dum.	4 – 6	0	6	6	10	10	4	4	
F	4 – 8	1	6	7	17	18	11	0	
Dum.	5 – 6	0	10	10	10	10	0	0	*
K	5 – 11	4	10	14	18	22	8	8	
G	6 – 7	3	10	13	10	13	0	0	*
H	7 – 9	5	13	18	13	18	0	0	*
Dum.	8 – 9	0	7	7	18	18	11	11	
L	8 – 11	3	7	10	19	22	12	12	
Dum.	9 – 10	0	18	18	20	20	2	0	
I	9 – 11	4	18	22	18	22	0	0	*
J	10 – 11	2	18	20	20	22	2	2	

* Critical

Exercise 9 (page 65)

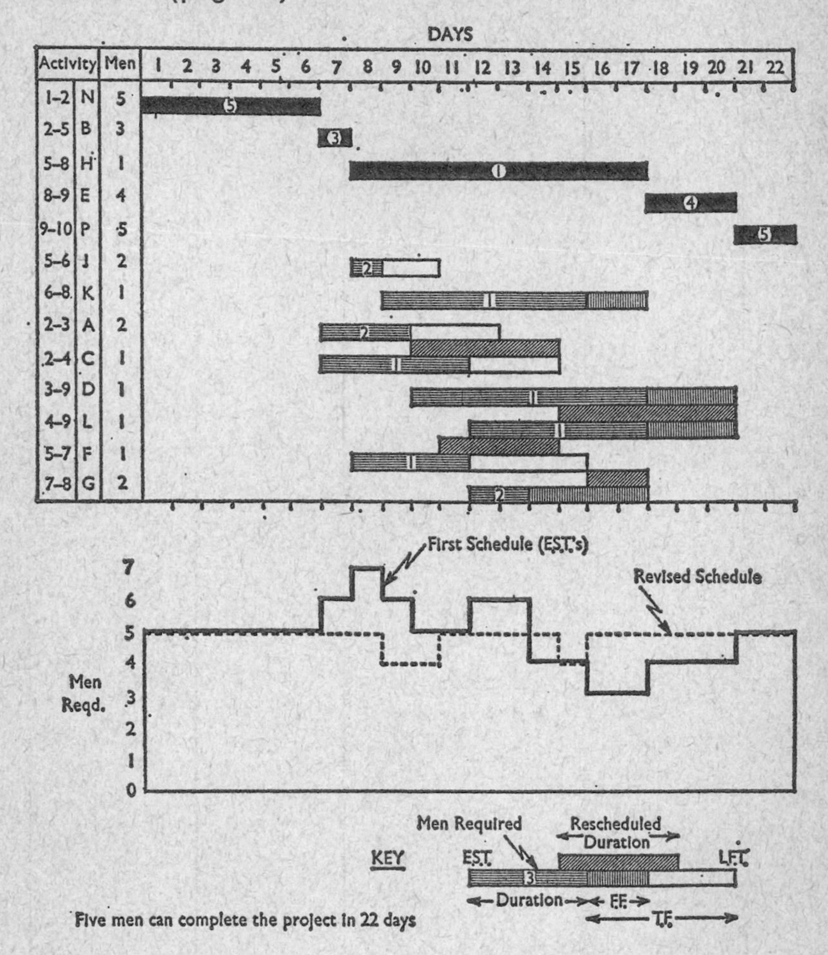

Five men can complete the project in 22 days

INDEX